On Sakharov

On Sakharov

EDITED BY ALEXANDER BABYONYSHEV
Translated by Guy Daniels

ALFRED A. KNOPF · NEW YORK · 1982

THIS IS A BORZOI BOOK
PUBLISHED BY ALFRED A. KNOPF, INC.

Translation Copyright © 1982 by Alfred A. Knopf, Inc.
ıll rights reserved under International and Pan-American Copyright Conventions.
Published in the United States by Alfred A. Knopf, Inc., New York,
and simultaneously in Canada by Random House of Canada Limited, Toronto.
Distributed by Random House, Inc., New York.

Originally published as *Sakharovskiĭ sbornik* by Khronika Press, New York.
Copyright © 1981 by A. Babyonyshev, R. Lert, and E. Pechuro.

LIBRARY OF CONGRESS CATALOGING IN PUBLICATION DATA
Sakharovskiĭ sbornik. English.
On Sakharov.

Translation of Sakharovskiĭ sbornik.
Translated essays by various Soviet writers and
scientists honoring Andrei Sakharov's sixtieth birthday.
1. Sakharov, Andreĭ Dmitrievich, 1921–   —Addresses, essays, lectures. 2. Civil
rights—Soviet Union—Addresses, essays, lectures. I. Babyonyshev, Alexander. II.
Sakharov, Andreĭ Dmitrievich, 1921–   . III. Title.
JC599.S58S2413   1982   323.4′092′4   81-48259
ISBN 0-394-52469-1   AACR2
ISBN 0-394-71004-5 (pbk.)

Manufactured in the United States of America
First Edition

Contents

II

From the Editor

This collection includes contributions from young people and old people, writers and scientists, believers and atheists—a cross-section from our motley times, which Vladimir Kornilov has called "the age of Sakharov."

Each one has contributed what he had in him, what broke forth at the thought of the man whose name resounds on every page. We did not try to avoid repetitions, and we are not ashamed of them: words of love given utterance are often similar. A great joy and gratitude, even though perhaps not too skillfully expressed, have brought into being these articles, poems, and letters. Gratitude for having encountered a man who is courageous and good, just, natural, and something of an oddity. Gratitude for the fact that he is living among us, that we are his contemporaries, and that someday our times will be measured by his stature.

ANDREI SAKHAROV

An Autobiographical Note

I was born on May 21, 1921, in Moscow. My father was a well-known physics teacher and the author of textbooks and popular science books. My childhood was spent in a large communal apartment most of whose rooms were occupied by our relatives, with only a few outsiders mixed in. Our home preserved the traditional atmosphere of a numerous and close-knit family—respect for hard work and ability, mutual aid, love for literature and science. My father played the piano well; his favorites were Chopin, Grieg, Beethoven, and Scriabin. During the Civil War he earned a living by playing the piano in a silent movie theater. I recall with particular fondness Maria Petrovna, my grandmother and the soul of our family, who died before World War II at the age of seventy-nine. Family influences were especially strong in my case, because I received my early schooling at home and then had difficulty relating to my own age group.

After graduating from high school with honors in 1938, I enrolled in the Physics Department of Moscow University. When war began, our classes were evacuated to Ashkhabad, where I graduated with honors in 1942. That summer I was assigned work for several weeks in Kovrov, and then I was employed on a logging operation in a remote settlement near Melekess. My first vivid impression of the life of workers and peasants dates from that difficult summer of 1942. In September I was sent to a large arms factory on the Volga, where I worked as an engineer until 1945.

I developed several inventions to improve quality-control procedures at that factory. (In my university years I did not manage to engage in original scientific work.) While still at the factory in 1944, I wrote several articles on theoretical physics, which I sent to Moscow for review. Those first articles have never been published, but they gave me the confidence in my powers which is essential for a scientist.

In 1945 I became a graduate student at the Lebedev Physical Institute. My adviser, the outstanding theoretical physicist Igor Tamm, who later became a member of the Academy of Sciences and a Nobel laureate, greatly influenced my career. In 1948 I was included in Tamm's research group, which developed a thermonuclear weapon. I spent the next twenty years continuously working in conditions of extraordinary tension and secrecy, at first in Moscow and then in a special research center. We were all convinced of the vital importance of our work for establishing a worldwide military equilibrium, and we were attracted by its scope.

In 1950 I collaborated with Igor Tamm in some of the first research on controlled thermonuclear reactions. We proposed principles for the magnetic thermal isolation of plasma. I also suggested as an immediate technical objective the use of a thermonuclear reactor to produce fissionable materials as fuel for atomic power plants. Research on controlled thermonuclear reactions is now receiving priority everywhere. The tokamak system, which is under intensive study in many countries, is most closely related to our early ideas.

In 1952 I initiated experimental work on magnetic-explosive generators (devices to transform the energy of a chemical or nuclear explosion into the energy of a magnetic field). A record magnetic field of 25 million gauss was achieved during these experiments in 1964.

In 1953 I was elected a member of the USSR Academy of Sciences.

My social and political views underwent a major evolution over the fifteen years from 1953 to 1968. In particular, my role in the development of thermonuclear weapons from 1953 to 1962, and in the preparation and execution of thermonuclear tests, led to an increased awareness of the moral problems engendered by such activities. In the late 1950s I began a campaign to halt or to limit the testing of nuclear weapons. This brought me into conflict first with Nikita Khrushchev in 1961, and then with the Minister of Medium Machine Building,[1] Efim Slavsky, in 1962. I helped to promote the 1963 Moscow treaty banning nuclear weapon tests in the atmosphere, in outer space, and under water. From 1964, when I spoke out on problems of biology,[2] and especially from 1967, I have been interested in an ever-expanding circle of questions. In 1967 I joined the Committee for Lake Baikal.[3] My first appeals for victims of repression date from 1966–67.

The time came in 1968 for the more detailed, public, and candid statement of my views contained in the essay "Thoughts on Progress, Peaceful Coexistence, and Intellectual Freedom."[4] These same ideas were echoed seven years later in the title of my Nobel lecture: "Peace, Progress, and Human Rights." I consider these themes of fundamental importance and closely interconnected. My 1968 essay was a turning point in my life. It quickly gained worldwide publicity. The Soviet press was silent for some time, and then began to refer to the essay very negatively. Many critics, even sympathetic ones, considered my ideas naive and impractical. But it seems to me, thirteen years later, that these ideas foreshadowed important new directions in world and Soviet politics.

[1]The Ministry of Medium Machine Building is responsible for nuclear weapons and industry in the USSR.—Trans.
[2]In 1964 Sakharov spoke out at the Academy of Sciences against political interference with biology and the persecution of geneticists during a debate on the election of one of Trofim Lysenko's associates.—Trans.
[3]The Committee for Lake Baikal was organized to protect Lake Baikal from industrial pollution; it was apparently sponsored by or at least tolerated by the authorities.—Trans.
[4]Sakharov's essay was first published in English by the *New York Times* (as "Progress, Coexistence, and Intellectual Freedom") and has been republished in *Sakharov Speaks* (New York: Knopf, 1974); an autobiographical note written by Sakharov in 1973 was published as an introduction to that volume.

After 1970, the defense of human rights and of victims of political repression became my first concern. My collaboration with Valery Chalidze and Andrei Tverdokhlebov,[5] and later with Igor Shafarevich[6] and Grigorii Podyapolsky,[7] on the Moscow Human Rights Committee was one expression of that concern. (Podyapolsky's untimely death in March 1976 was a tragedy.)

After my essay was published abroad in July 1968, I was barred from secret work and excommunicated from many privileges of the Soviet establishment. The pressure on me, my family, and my friends increased in 1972, but as I came to learn more about the spreading repressions, I felt obliged to speak out almost daily in defense of one victim or another. In recent years I have continued to speak out as well on peace and disarmament, on freedom of association, movement, information, and opinion, against capital punishment, on protection of the environment, and on nuclear power plants.

In 1975 I was awarded the Nobel Peace Prize. This was a great honor for me, as well as recognition for the entire human rights movement in the USSR. In January 1980 I was deprived of all my official Soviet awards (the Order of Lenin, three times Hero of Socialist Labor, the Lenin Prize, the State Prize) and banished to Gorky, where I am virtually isolated and watched day and night by a policeman at my door. The regime's action lacks any legal basis. It is one more example of the intensified political repression that has gripped our country in recent years.

Since the summer of 1969 I have been a senior scientist at the Academy of Sciences' Institute of Physics. My current scientific interests are elementary particles, gravitation, and cosmology.

I am not a professional politician. Perhaps that is why I am always bothered by questions concerning the usefulness and eventual results of my actions. I am inclined to believe that moral criteria in combination with unrestricted inquiry provide the only possible compass

---

[5]Valery Chalidze and Andrei Tverdokhlebov are physicists who have left the USSR under pressure from the authorities.—Trans.
[6]Igor Shafarevich is a Moscow mathematician who is a corresponding member of the Academy of Sciences.—Trans.
[7]Grigorii Podyapolsky was a geophysicist.—Trans.

for these complex and contradictory problems. I shall refrain from specific predictions, but today as always I believe in the power of reason and the human spirit.

ANDREI SAKHAROV

Gorky, March 24, 1981

Some Events in the Scientific and Public Careers of Andrei D. Sakharov

1942

Graduated from Moscow State University and was sent to work as an engineer at a war plant.

1943–44

Made a number of inventions for facilitating increased production. In particular, proposed a simple method for determining the thickness of nonmagnetic coatings of bullets, and a device for detecting inadequately tempered cores of armor-piercing shells, which eliminated the necessity of sample testing (was awarded a certificate of authorship for this invention).

Wrote his first scientific papers.

1945

Enrolled as a graduate student in the P. N. Lebedev Physical Institute of the Academy of Sciences of the USSR.

1947

Defended his thesis for the status of doctoral candidate, on the topic "Theory of 0–0 Nuclear transitions," ZhETF, 1948.

Article: "The Production of Mesons" (the editors changed the title to "The Production of the Hard Component of Cosmic Rays"), ZhETF, Vol. 17.

1948

Article: "The Excitation Temperature of a Gas-Discharge Plasma," Izvestiya Akad. Nauk SSSR, Vol. 12.

Article: "The Interaction Between Electron and Positron in Pair Production," ZhETF, Vol. 18.

Developed the theory of μ-mesonic catalysis (together with Ya. B. Zeldovich). Published in ZhETF, 1957 [Sov. Phys. JETP 5, 775 (1957)].

Included in research group for the development of thermonuclear weapons.

1950–51

Formulated the principles of a controlled thermonuclear reaction based on magnetic thermal isolation of a high-temperature plasma (together with I. E. Tamm). The results were reported by I. Kurchatov at a conference at the Harwell Laboratory and were published in the proceedings of the First Geneva Conference on the Peaceful Uses of Nuclear Energy. This principle was the basis of the project for a controlled reactor called tokamak, now being intensively developed in the USSR and other countries.

1951–52

Proposed the principles of production of ultrastrong magnetic fields by the use of the energy of explosions and of the construction of magnetoplosive generators.

1953

Awarded the academic degree of Doctor of Physical and Mathematical Sciences. Elected a member of the USSR Academy of Sciences. Awarded the Order of Lenin. Awarded the title of Hero of Socialist Labor and a State (Stalin) Prize.

1955

Confrontation with Marshal M. Nedelin. (Sakharov expressed the hope that the nuclear weapon being tested would never be used, and the marshal responded with a parable, the gist of which was that in deciding such questions the leaders can get along without advisers.)

1956

Awarded the title of Hero of Socialist Labor for the second time. Awarded a Lenin Prize.

1957

Article on the danger of nuclear testing in a scientific journal (reprinted in the magazine *The Soviet Union* and in many foreign publications).

1958

A talk with M. A. Suslov, secretary of the CC CPSU, about the unfavorable situation in biology, and about the fate of the unjustifiably arrested Dr. I. Barenblat, on which subject Sakharov had written to the Central Committee. (Barenblat was soon released.)

Jointly with I. Kurchatov, spoke out against planned nuclear tests. (Kurchatov made a special plane trip to Yalta to see Khrushchev, but did not succeed in stopping the tests.)

Article: "Theory of a Magnetic Thermonuclear Reactor," with I. E. Tamm, in collection *Plasma Physics and the Problems of Controlled Thermonuclear Reactions,* Vol. 1, Part 2.

1961

Note to Khrushchev at a meeting of leaders and atomic scientists on the necessity of keeping a moratorium on nuclear testing. In his speech at the banquet, Khrushchev replied that political decisions, including the matter of testing nuclear weapons, were the prerogative of the party and government leaders and did not concern scientists.

Proposal for utilizing the heating of deuterium with a beam from a pulsed laser to obtain a controlled nuclear reaction. (At the present time this principle is being actively developed in the USSR and the U.S. as one of the most promising ways of obtaining a controlled thermonuclear reaction.)

1962

Awarded the title of Hero of Socialist Labor for the third time.

Conflict with the Ministry of Medium Machine Building on the testing of a nuclear weapon of great capacity, useless from the scientific and technical viewpoint and threatening the lives of many people. Futile appeal to Khrushchev with a view to stopping the planned testing.

Proposal on concluding an agreement to ban nuclear testing in the atmosphere, under the water, and in space. (Approved by the higher Soviet leadership and put forth in the name of the USSR. In 1963 Khrushchev and Kennedy signed the "Moscow agreement" on banning nuclear testing in the three environments, which was signed by the majority of states.)

1964

Speech at the elections at the USSR Academy of Sciences against the candidacy of T. Lysenko's close collaborator, N. Nuzhdin. (Nuzhdin was not elected. Lysenko, president of VASKhNIL— the V. I. Lenin All-Union Academy of Agricultural Sciences—published a newspaper article attacking "the engineer Sakharov.")

Letter to Khrushchev on the situation in biological science. (This letter was mentioned by Suslov among many charges brought against Khrushchev following his removal.)

1965

Article: "Magnetic Cumulation," with R. Z. Lyudaev, E. N. Smirnov, Yu. N. Plyushchev, A. I. Pavlovskii, V. K. Chernishev, E. A. Feoktistova, E. I. Zharinov, and Yu. A. Zysin, Dokl. AN SSSR, *165*, No. 1 (1965) [Soviet Phys. Doklady *10*, 1045 (1965)].

Article: "The Initial Stage of the Expansion of the Universe and the Appearance of Nonuniformities in the Distribution of Matter," ZhETF *49*, 345 (1965) [Sov. Phys. JETP *22*, 241 (1965)].

1966

Appeal, together with other well-known scientists, artists, and writers (twenty-two persons in all), to the 23rd Congress of the CPSU against attempts to rehabilitate Joseph Stalin.

Telegram to the RSFSR Supreme Soviet protesting against the introduction of Article 190-1 ("circulating maliciously false and slanderous fabrications defaming the Soviet state and social system") as a pretext for prosecuting people for their convictions.

First participation in a demonstration at the Pushkin Monument (yearly demonstrations on Constitution Day for human rights and against the unconstitutional articles of the Criminal Code).

Article: "Magnetoplosive Generators," Usp. Fiz. Nauk 88, No. 4 (1965) [Sov. Phys. Uspekhi 9, 294 (1966)].

Article: "Quark Structure and Masses of Strongly Interacting Particles," with Ya. B. Zeldovich, Yad. Fiz. 4, 395 (1966) [Sov. J. Nucl. Phys. 4, 283 (1966)].

Article: "On the Maximum Temperature of Thermal Radiation," Pis'ma ZhETF 3, 439 (1966) [JETP Lett. 3, 288 (1966)].

1967

Letter to L. Brezhnev in defense of A. Ginzburg, Yu. Galanskov, V. Lashkova, and Yu. Dobrovolsky.

Participation in the work of the committee on the problem of Baikal. A talk with Brezhnev on Baikal. (Sakharov and certain other scientists actively struggling to save Lake Baikal were not invited to the session at which the question of using the lake was decided. With the approval of M. Keldysh, speaking in the name of the Academy of Sciences, a decision was made to build a cellulose-paper plant on Lake Baikal.)

Publication of an article in *Budushcheye Nauki* ("The Future of Science"), with a prognosis of the development of science and technology (not put on sale).

Article: "Violation of CP Invariance, C Asymmetry, and the Baryon Asymmetry of the Universe," Pis'ma ZhETF 5, 32 (1967) [JETP Lett. 5, 24 (1967)]. This is a pioneering paper which anticipated the main current of research in this field for ten to twelve years. It is referred to by practically everyone who has occupied himself with the asymmetry of the universe and the instability of the proton—a question of key significance in current theorizing about the evolution of the universe.

Article: "Quark-Muon Currents and the Violation of CP Invariance," Pis'ma ZhETF 5, 36 (1967) [JETP Lett. 5, 27 (1967)].

Article: "Vacuum Quantum Fluctuations in Curved Space and the Theory of Gravitation," Dokl. AN SSSR *177*, 70 (1967) [Sov. Phys. Doklady *12*, 1040 (1968)].

Article: "Vacuum Polarization and the Theory of the Zero Lagrangian of the Gravitational Field." Preprint, Applied Math. Section of the Steklov Math. Inst., Acad Sci. USSR.

1968

Article: "Thoughts on Progress, Peaceful Coexistence, and Intellectual Freedom." The article was widely distributed in samizdat, and in July of 1968 was published abroad. The year 1968–69 witnessed the publication of about forty editions, with an overall printing of more than 18 million copies.

Barred from secret work, owing to the publication of "Thoughts. . ."

Gave almost all of his personal savings to the Red Cross and for the construction of a cancer hospital.

1969

Elected a foreign member of the Boston Academy of Science and Arts. (In subsequent years was elected foreign member of the U.S. National Academy of Science and the New York Academy, received honorary doctorates from Siena, Jerusalem, and other universities, and was made an honorary citizen of Florence and Turin.)

Went to work in the Theoretical Physics Department of the Lebedev Physical Institute of the USSR Academy of Sciences.

Article: "Antiquarks in the Universe," in collection *Problems of Theoretical Physics*.

Article: "A Many-Sheet Model of the Universe," Preprint Appl. Math. Sec. Steklov Math. Inst., Acad. Sci. USSR.

1970

Letter to the CC CPSU, USSR Council of Ministers, and Presidium of the USSR Supreme Soviet (jointly with V. Turchin and R. Medvedev) on the necessity for democratizing society for the development of science, the economy, and culture.

Collection of signatures on an appeal in defense of Zhores Medvedev, confined in a psychiatric hospital.

Present at the trial of the mathematician R. Pimenov and the actor B. Vail, accused of distributing samizdat.

Founding of the Human Rights Committee (A. Sakharov, V. Chalidze, A. Tverdokhlebov, and, later, G. Podyapolsky and I. Shafarevich). The committee considered and adopted appeals on several important problems; in particular, on forcible psychiatric hospitalization for political reasons, and on the forcible resettlement of individuals and peoples.

Spoke out for commuting the death sentences of E. Kuznetsov and M. Dymshits, and reducing the sentences of the other defendants in the "airplane trial."

1971

Sent Brezhnev a "Memorandum" on urgent problems of domestic and foreign policy. (No answer was received.)

Appeal on the subject of persons confined in special psychiatric hospitals.

Letter to N. Shchelokov, Minister of Internal Affairs, on the situation of the Crimean Tatars. A talk at the Ministry of Internal Affairs.

Open appeal to members of the Presidium of the USSR Supreme Soviet on freedom of emigration and unobstructed return.

1972

"Memorandum" and an "Afterword" to it published abroad.

Drafting of appeals to the USSR Supreme Soviet on amnestying political prisoners and abolishing capital punishment. Collecting signatures on these documents.

Participation in the demonstration at the Lebanese Embassy protesting the murder of the Israeli athletes at the Munich Olympics.

First interview with foreign correspondents published in the West. (Given in connection with the trial of the astrophysicist K. Lyubarsky.)

Participation in the work of the editorial committee in preparing the collection *Problems of Theoretical Physics*, dedicated to I. E. Tamm.

Article: "The Topological Structure of Elementary Charges and CPT Symmetry," in the collection *Problems of Theoretical Physics*.

1973

Interview with a correspondent of the Swedish radio on the political, economic, and social problems facing the country.

Press conference on the dangers of unilateral détente. Also, a talk with Malyarov, USSR Deputy General Procurator, and a warning from the Procurator's Office. (At the same time the Soviet press launched a vicious campaign against Sakharov. He was attacked, collectively and individually, by writers, composers, workers, and scientists, in particular by a large group of academicians. Members of his family were also subjected to attacks in the press and to various kinds of harassment. His wife, Elena Bonner, was called in for interrogation by the KGB several times. Sakharov responded to these threats and harassments by explaining his position to Western correspondents in an interview.)

Visit to Sakharov's apartment by people claiming to be members of the Palestinian terrorist organization Black September, who threatened to kill Sakharov and his family, and demanded that he disown his statement "On the October War in Israel."

Autobiographical foreword to the collection of articles and statements *Sakharov Speaks*.

Awarded a prize by the International League for Human Rights.

1974

Interview and article dealing with Aleksandr Solzhenitsyn's book *The Gulag Archipelago*.

Article: "On Aleksandr Solzhenitsyn's 'Letter to the Leaders of the Soviet Union.'" (The danger of isolationist and nationalist tendencies is noted.)

Futurological article: "The World Fifty Years from Now" (published in *Saturday Review*).

Hunger strike with a demand for the release of political prisoners, timed to coincide with Nixon's visit to the USSR.

Appeal to the U.S. Congress on the subject of the Jackson Amendment.

Transmittal to the Chancellor of the Federal Republic of Germany of a list of 6,000 Germans in Kazakhstan desirous of emigrating.

Awarded a prize by Freedom House in the U.S.

Article: "On the Scalar-Tensor Theory of Gravitation and the Zero-Lagrangian Hypothesis," Pis'ma ZhETF 20, 189 (1974) [JETP Lett. 20, 81 (1974)].

1975

Appeal (jointly with the writer Heinrich Böll) for the amnestying of political prisoners.

Second appeal for putting a stop to genocide in Iraqi Kurdistan. (The first was made in late 1974.)

Appeal to Suharto, President of Indonesia, for the amnestying of political prisoners.

Letter to the Pugwash Conference on disarmament, peace, and international trust.

Awarded the Nobel Peace Prize.

The first Sakharov Hearings (a regular international seminar considering problems of human rights in the USSR and the countries of Eastern Europe) are held in Copenhagen.

Trip to Vilnius for the trial of Sakharov's friend, the biologist Sergei Kovalev.

Nobel lecture, "Peace, Progress, and Human Rights," read by Elena Bonner on December 11 after receipt of the prize at the Swedish Academy of Science.

Autobiography for the Nobel collection is written.

Article: "The Spectral Density of Eigenvalues of the Wave Equation and Vacuum Polarization," Teor. Mat. Fiz. 23, 178 (1975) [Theor. Math. Phys. (USSR) 23, 435 (1975)].

Article: "Mass Formula for Mesons and Baryons with Effects of Charm Included," Pis'ma ZhETF 21, 554 (1975) [JETP Lett. 21, 258 (1978)].

1976

Speech at the funeral of G. Podyapolsky; foreword to his book is written.

Trip to Omsk for the trial of R. Dzhemilev.

Elected vice-president of the International League for Human Rights.

Appeal (jointly with Yu. Orlov and V. Turchin) to the conference of leaders of the European Communist parties proposing that the question of human rights be included in the agenda of the conference.

Appeal (jointly with Elena Bonner) to the UN concerning the tragic situation in the Palestinian camp of Tel-Zaatar.

Appeal to the Committee in Defense of Polish Workers.

Participation in the symposium "Jewish Culture in the USSR."

1977

Appeal to U.S. President Jimmy Carter in defense of P. Ruban.

Appeal to world public opinion concerning attempts to blame the dissidents for explosions in the Moscow subway. (This appeal provoked a second warning to Sakharov from the USSR Procurator's Office, which was followed by harassment of his family.)

Exchange of letters with President Carter.

Article "Alarm and Hope" in a collection of articles by winners of the Nobel Peace Prize.

Article against capital punishment for the Amnesty International symposium in Stockholm.

Letter to the Steering Committee of the Sakharov Hearings in Rome.

Article: "Nuclear Power and the Freedom of the West."

Statement on the amnesty decree issued by the Presidium of the USSR Supreme Soviet (a demand that amnesty be extended to political prisoners).

Appeal (jointly with Elena Bonner) to President Tito concerning amnesty for prisoners in Yugoslavia.

Speech read at a congress of the AFL-CIO, read from a transmitted text.

Trip (accompanied by Elena Bonner and A. Semenov) to the Mordovian camps for a visit with E. Kuznetsov. After eleven days of waiting, the visit was disallowed.

Awarded the International Joseph Prize (Anti-Defamation League).

1978

Confrontation with police during the trial of Yu. Orlov. (Sakharov and his wife were detained and fined for having allegedly disturbed public order.)

Article: "The Human Rights Movement in the USSR and Eastern Europe: Its Goals, Significance, and Difficulties."

Letters to Brezhnev (during his visit to West Germany) and Helmut Schmidt concerning the arrest of the worker I. Wagner, accused of parasitism. (Wagner was released.)

"Afterword" to the collection of Sakharov's writings *Alarm and Hope* (published in New York).

1979

Letter to Brezhnev concerning Decree No. 700 of the Council of Ministers (on the Crimean Tatars). (In April, Sakharov had transmitted to the French Embassy a letter to Valéry Giscard d'Estaing from a group of Crimean Tatars.)

Letter to Brezhnev requesting postponement of execution and a public trial of Zatikyan, Bagdasaryan, and Stepanyan, accused of having caused an explosion in the subway. Wrote a foreword to Malva Landa's article about this trial. Sakharov's apartment visited by people who call themselves relatives of those killed in the explosion and who threaten him.

Trips to Tashkent for the trials of M. Dzhemilev and V. Shelkov.

Drafting of the text of a speech to be given at the New York Academy of Science in connection with the awarding of a prize by the academy.

Review of Freeman Dyson's book *Disturbing the Universe*, which appeared in *Washington Post Book World*.

Text of a speech to be given at the Sakharov Hearings in Washington.

Appeal to Brezhnev on the problem of the unobstructed delivery of foodstuffs to Cambodia.

Appeal in connection with the conviction and sentencing of members of the human rights movement Charter-77.

Article: "The Baryon Asymmetry of the Universe," ZhETF 76,
1172 (1979) [Sov. Phys. JETP 52, 349 (1979)]. (Development of ideas
stated in a paper in 1976, from the point of view of current models of
elementary particles.)

1980

Interview with Western correspondents about the sending of Soviet
troops into Afghanistan.

Arrested on the street January 22 and taken to the USSR Procur-
ator's Office. Deputy General Procurator read aloud the decree of the
Presidium of the USSR Supreme Soviet dated January 8 depriving A.
Sakharov of his government awards and prizes.

Taken to Gorky, where to the present day he is living in almost
total isolation, under constant guard by the KGB and the police.

Statement as to the illegality of the repressions undertaken,
with a demand for judicial examination of the charges brought against
him.

Several interviews by letter with foreign journalists.

Article: "An Alarming Time," published by *The New York Times
Magazine*.

A statement in connection with the conviction and sentencing of
V. Nekipelov, T. Velikanova, V. Stus, and other repressed persons.
A statement on the case of Liza Alekseyeva.

Letter to Brezhnev about Afghanistan.

Letter to A. Aleksandrov, president of the USSR Academy of Sciences, stating his position on basic political problems. (Censure of the Academy of Sciences for refusing to
defend scientists subjected to repression, including Sakharov
himself.)

Elected a foreign member of the Accadèmia dei Lincei at Rome.

Article: "Mass Formula for Mesons and Baryons," ZhETF 79, 2112
(1980) [Sov. Phys. JETP 52, 1059 (1980)].

Article: "Cosmological Models of the Universe with Reversal of
Time's Arrow," ZhETF 79, 698 (1980) [Sov. Phys. JETP 52, 349
(1980)].

Article: "Estimate of the Constant Interaction of Quarks with the
Gluon Field," ZhETF 79, 350 (1980) [Sov. Phys. JETP 52, 175 (1980)].

1981

Elected a foreign member of the French Academy of Sciences. (Documents sent to the Presidium of the Academy of Sciences were not forwarded to Sakharov.)

Statement to the press in connection with the theft of his manuscripts, diary, letters, and materials for scientific articles.

Appeal (jointly with Elena Bonner) in connection with the arrest of A. Marchenko.

Article: "The Responsibility of Scientists," written for the conference organized by the New York Academy on the occasion of Sakharov's sixtieth birthday.

Article: "How to Preserve World Peace," written for *Parade* magazine.

Personal appeal to Leonid I. Brezhnev to grant an exit visa to Liza Alekseyeva, who for several years had been denied permission to join her husband, Sakharov's stepson, in the United States—in Sakharov's view, becauuse of his own public activities.

A "Letter to My Foreign Colleagues," in which Sakharov explains the reasons for the hunger strike that he and his wife have decided to begin.

November 22: The hunger strike begins.

December 4: Sakharov and his wife are forcibly hospitalized.

December 8: The hunger strike is discontinued, after Sakharov is informed by KGB officials that his daughter-in-law will be allowed to leave the country.

Appeal to the Communist party leader of Soviet Georgia, Edward A. Shevarnadze, in connection with a new arrest of the Georgian human rights activist Merab Kostava.

1982

Appeal to President Mitterrand of France on behalf of Anatoly Shcharansky. Statements in defense of I. Kovalev, G. Vladimov, and S. Kalistratova. A "Letter to Soviet Scientists," in which Sakharov urges them to speak up on social issues and in defense of their persecuted colleagues.

Letter to the Pugwash Conference.

Article, "Multifaceted Models of the Universe," scheduled to appear in ZhETF (Sov. Phys. JETP).

Dear Andrei Dmitriyevich:

Your sixtieth birthday has been cast into gloom by the hard lot of your friends, by the illegality of your exile, by the constant watch kept at your door. They have deprived you of your government awards, they have deprived you of scientific and personal contacts. They have taken away everything that constituted the essence of your life: diaries, reflections about the past and the future, scientific projects. But no one has the power to deprive you of your incomparable righteousness and of our "nongovernmental" love for you. The 21st of May is celebrated in our hearts as a holiday—a holiday of reason, of the good, of spiritual greatness. Russia has shown the world its latent strength through your feat, repeated every day. As Leo Tolstoy pointed out, spiritual strength can be suppressed only until "it reaches the highest stage, at which it is more powerful than everything." The spiritual force radiated by you is growing, and cannot be taken away along with your papers. Your words stir people to active goodness. Your thought agitates and alarms hearts, takes possession of thousands, whether free or behind bars, teaches people to think and leads them from one stage of consciousness to the next. On the holiday of your sixtieth birthday I send you this wish: May moral strength pre-

3

vail over crude force; may the treasures taken from you be returned to you; and may you and all unjustly persecuted people soon return home.

LIDIA CHUKOVSKAYA

VLADIMIR KORNILOV

Evenings in the Kitchen

Evenings in the kitchen
At Andrei
Dmitriyevich's, evenings in the kitchen . . .
Although winter, fiercely chilling,
Had begun its age-old work,

Evenings in the Sakharovs' kitchen
Continued even in winter,
And hopes had not yet died out,
And we sat shoulder to shoulder.

I was happy. I watched with eyes
Full of rapture and love
As our host kept quiet and listened to us,
Not imposing his own views.

Only on his forehead—trying to conceal itself—
A childlike, shy bravery
And the lofty sorrow of the soul
Showed as if by chance.

.   .   .

5

I went out of my mind, all of me straightening up;
At last I had met, at last,
The prototype of true democracy,
The model of equality and brotherhood!

But it never occurred to him—
I could see that clearly—to take precedence.
And not a fire-breathing prophet.

. . . But on the other hand, like Delvigs and Kyukhls,[1]
Who belonged to the age of Pushkin,
All those who were in Sakharov's kitchen,
In the old days, if only at dawn,

All those not yet locked up in camps,
All those in exile, half locked up,
All those who are teaching peace from New York professorial chairs,
Or going blind in the BURs of Perm,[2]

Like the words in one poem,
Are irrevocably included
In the age named after Sakharov,
The best in the country's history.

[1]Anton Delvig and Wilhelm Küchelbecker were minor poets who were close friends of Pushkin's.—Trans.
[2]BUR: a strict-discipline barracks in a labor camp. Perm is an industrial city in the Urals in whose vicinity several camps for political prisoners are located.—Trans.

In Exiling Him to Gorky

In exiling him to Gorky without an investigation or trial, without announcing a verdict or sentence, in taking this most extraordinary measure, the regime accorded him an honor such as one would expect to be granted only to a hereditary prince or a potential president. And yet he hardly claimed to possess any material power; he didn't serve as the leader of anything; he didn't head up a party or an organization or even the handful of his kindred spirits. He was only the most brilliant spokesman for that incorporeal power that is called "moral resistance." Is that a lot or a little? The regime decided that perhaps it was a second government.

From time to time they write or record on tape their tongue-tied memoirs. There is hope, therefore, that someday we'll learn the "level" at which the decision was taken, what the pros and cons were, and who dared to sign. I think they'll have a lot of trouble evaluating whether they won or lost in trying to separate Andrei Dmitriyevich Sakharov from the human rights movement in Russia. It's hardly likely they'll come to the realization that this didn't depend upon them at all.

Today, as he turns sixty, one can definitely say that Andrei Sakharov is without doubt the greatest success of the democratic movement, the embodiment of its conscience, the justification for all its mistakes and defeats. In his beautiful maturity he is a star of the first

magnitude on the horizon of our social life. This is recognized by both
his friends and his enemies, and by those who for lack of decisiveness
belong to neither category.

I have had the good fortune to know him for almost ten years,
and have never met a more impressive or humane person. It is possible, barring unforeseen circumstances, that I will never see him
again. But I would like to wish him—in addition to those good things
we usually wish for a person on his birthday—I would like to wish
that that freedom which is our birthright, and his more than anyone
else's, should come to him not through the capricious mercy of his
oppressors but through an unexpected and mighty turning of history
before which they will be rendered powerless.

MARIA PETRENKO-PODYAPOLSKAYA

In the Thick of Events

Almost out of breath, I ran past the crossings. What was pounding in my head was not thoughts but an influx of anxieties, spreading throughout my body and tensing my muscles. Along the way I stopped at pay phones, trying to call. Kursky. Under the ground, passing under the Sadovoye. I thought: Maybe go by trolley bus beyond the Yauza to see what's going on at the building? But I'd have to wait for the trolley bus, so I ran.

From the hill I saw a crowd at the entrance. (As it turned out later, they were correspondents.) Just beyond Obukh Lane a character with a little red book in his hands showed up and asked me for my passport. Another character was looking over his shoulder. They put me in a car and took me a short distance away, almost across the street, to a building on the Yauza. It was plain to see that the yard there had formerly been a park. Then it had been a children's tuberculosis sanatorium, and now it was a police station. I sat there in the Red Corner[1] for four hours. They weren't interested in me, but they didn't let me go. I felt dull, as if in a railroad station, when waiting is inevitable.

I realized that this day marked a new era. Until today they had

[1]A room where meetings and other events of ideological or ritual significance are held.—Trans.

9

not touched Andrei, but now they had decided. What was it? Were
they confident that they could get away with it? Were they testing to
see how the world would react? Was this again the complete lawlessness of Stalinist times? Was that really possible today? How would it
all end? It was frightful to think of.

Then some policeman came in and said, "You can go." At the door
I ran into Slava Bakhmin.[2] It turned out that they had detained him
too. The two of us went across the street, and were joined at the
entrance by Felix Serebrov.[3] At the doorway to the apartment two
policemen in full-dress uniform (white straps and armlets) stunned us
with the news: "Everybody has gone to the airport to see him off."
That's all we could find out. Either they weren't "supposed" to say
anything more, or they didn't know anything more. Or maybe what
they said was fake.

Thus did I remember that black day. One of the people most
respected by mankind, Academician Andrei Sakharov, winner of the
Nobel Peace Prize and a great Russian physicist, was shipped off to
Gorky without an investigation or a trial. In Gorky he was assigned
living quarters in an apartment guarded by the police and the KGB
where he is isolated, forbidden correspondence, etc., etc.

Throughout the next few days we listened to the radio. It seemed
that the world outside of our country was ready to burst with indignation. Those protesting included governments, social organizations
and associations, political and other public figures, scientists, writers,
and simply people who had known him personally or known of his
activities.

We, his compatriots, protested too. Not all of us, alas, but only
those who in their own cases regarded keeping silent as a form of
collaborating with the regime. But then I'm probably not right; among
those who kept silent were some who suffered from their silence but
could not overcome their great fear of the repressive machine. The

[2]Vyacheslav Ivanovich Bakhmin, mathematician, member of the Working Commission
to Investigate the Misuse of Psychiatry for Political Purposes of the Moscow Group to
Promote Observance of the Helsinki Accords in the USSR. Arrested in May 1980 and
sentenced to three years in labor camp.
[3]Felix Arkadevich Serebrov, member of the Moscow Group to Promote Observance of
the Helsinki Accords in the USSR and of the Working Commission to Investigate the
Misuse of Psychiatry for Political Purposes. Now in a remand prison.

lessons of the past had been reinforced by lessons of the present, and that paralyzed many people. Among those in this category who kept silent were almost all of the scientists (including the academicians) and all scientific and social organizations, including the Academy of Sciences. This sad and shameful circumstance ensured the success of the undertaking. Andrei is still in Gorky; the regimen of his confinement is gradually being made harsher; and of those who walked hand in hand with him in the human rights movement, few are still free. The months before his exile witnessed the arrest of Tatyana Velikanova,[4] Victor Nekipelov,[5] and Malva Landa.[6] Shortly after his exile came the arrests of Alexander Lavut,[7] Leonard Ternovsky,[8] Vyacheslav Bakhmin, Felix Serebrov, Tatyana Osipova,[9] and many others. A wave of repressions rolled across the whole country. Tyranny now rages in the camps. People under investigation and prisoners are told by officials: "Now it's all over. You don't have your defender anymore."

This day in the history of our country looks frightful and irrepara-

---

[4]Tatyana Mikhailovna Velikanova, mathematician, member of the Initiative Group for the Defense of Human Rights in the USSR. Arrested in November 1979, sentenced under Article 70 of the RSFSR Criminal Code to five years in a strict-regimen camp and four years of exile.

[5]Victor Aleksandrovich Nekipelov, physician, poet, and publicist. Member of the Moscow Group to Promote Observance of the Helsinki Accords in the USSR. Arrested on December 7, 1979, and sentenced to seven years in a strict-regimen camp and five years of exile. Had previously served two years in a general-regimen camp after sentencing under Article 190-2.

[6]Malva Noyevna Landa, geologist, member of the Moscow Group to Promote Observance of the Helsinki Accords in the USSR. Arrested in July 1979 and sentenced to five years of exile. In 1977 she had been exiled for two years under the pretext of "a fire in her apartment."

[7]Alexander Pavlovich Lavut, mathematician. Member of the Initiative Group for the Defense of Human Rights in the USSR. Arrested in April 1980. Sentenced under Article 190 to three years in a general-regimen camp.

[8]Leonard Borisovich Ternovsky, radiologist. Member of the Moscow Group to Promote Observance of the Helsinki Accords in the USSR, and of its Working Commission to Investigate the Misuse of Psychiatry for Political Purposes. Arrested in April 1980, sentenced under Article 190-1 to three years in a general-regimen camp.

[9]Tatyana Semenovna Osipova, philologist. Member of the Moscow Group to Promote Observance of the Helsinki Accords in the USSR. Arrested in May 1980 and sentenced by the Moscow City Court to five years in a general-regimen camp and five years of exile.

ble. There is only one hope: This is only today, and no one knows
what will come tomorrow.

My husband, Grigorii Podyapolsky, and I first heard about Andrei
Sakharov after the appearance in samizdat of his article "Thoughts on
Progress, Peaceful Coexistence, and Intellectual Freedom." It produced a very deep impression. People were amazed by the breadth
of the author's position, so uncharacteristic of the official—and even
of the samizdat—literature of our society. As always, the world proved
to be small. We had mutual acquaintances and contacts that went
back to the past. By itself, as it were, information about him emerged.
He was one of the creators of the hydrogen bomb, but was himself
opposed to nuclear tests. He protested vigorously against the pollution of the environment. Andrei Sakharov came from the same circle
of the Moscow intelligentsia to which the forebears of Grisha Podyapolsky had belonged, the same one to which our families now belong.
    We met for the first time at the home of Valery Chalidze. Very
soon Andrei, his wife, Elena Bonner, and her mother, Ruf Grigoryevna, along with the children, became an integral part of our life.
There was the joy of recognition, and many shared experiences.
    After the era of Stalinist thought control and the brief "thaw,"
many caring, active people tried to find a path for the development
of the fatherland, tried to bring the bitter experience of the past to
the broad masses. They hoped that their activity would help the country
to change over from despotism to democracy, and hence to a flower-
ing. In 1970, Valery Chalidze, Andrei Sakharov, and Andrei Tverdokhlebov formed the Human Rights Committee, which was later
joined by Igor Shafarevich and Grigorii Podyapolsky. *Khodoki*[10] kept
coming from all over the country to the Sakharovs' apartment on
Chkalov Street. It seemed to all of them that if Sakharov, thrice a
Hero of Labor and the winner of state prizes, took up their cause,
justice would triumph. He wrote, requested, and protested; and at
first it was effective, but later on it was not. However, the thirst for

---

[10]Roughly, "petitioners." Originally, a *khodok* ("walker") was a peasant chosen by his
community to travel to see an influential person and lodge a petition with him.

justice is so great among the people that even after Sakharov was exiled to Gorky the "stream of *khodoki*" did not dry up.

Of special importance for our time, and for future times, are Andrei's statements on basic problems of peace, progress, and human rights. However difficult the circumstances in which he has been put by the regime and by life itself, his voice is always free, reasonable, and clear. Much of what I have experienced as feeling, emotions, acquires in his formulations the sense of genuine problems without whose solution the world could not get along.

I am not a master of eloquence. My admiration for the personality of Andrei Sakharov is almost boundless. As a matter of everyday experience, I simply love him, as I love the whole Sakharov-Bonner family, and I would like to reminisce. For what is left to us from the past except memories . . . and experience?

A clear autumn morning. A Sunday in October 1973. The three of us—my husband, Grisha, Tanya Khodorovich, and I—are going to the Sakharovs' apartment on Chkalov Street. My husband and Tanya are to discuss something and make a decision, and I'm going along with them. That day, Grisha was going off on a business trip. It had been agreed in advance that we would come, so they must be expecting us. We ring at the door. We can hear voices from within the apartment, some kind of noise, and wrangling. . . . We listen. A chilling thought: Could it be a search? Grisha and Tanya remain at the door, while I run to a phone booth to make a call to the apartment. I return with a report that all I heard was a long hum; no one came to the phone.

The noise in the apartment has quieted down. A celebration seems to be in progress in the apartment above, on the eighth floor, and voices can be heard. Was the wrangling going on there? No. We very clearly heard voices from behind the door. We decide that I should go to the telephone booth again and call the children in Novogireyevo and somebody else, at my discretion, while Grisha and Tanya continue to ring the doorbell.

I am amazed that there are no "dawdlers" around the building, and apparently there were none when we came. That's unusual.

The pay phone near the building is not working. That's usual. I have to run to the Yauza, where I find a phone that works. After calling several numbers, I begin to shake. Something's going on. I run back. Near the Yauza I meet up with Grisha and Tanya. They are alarmed because I was gone so long. They tell me nothing has changed: no sign of life from behind the door.

We go back to the apartment building and see that taxis are driving up and people we know are getting out. My calls have yielded results. We hurry upstairs. Again, the door is closed. Where did the new arrivals go? We ring, and the door is opened. The apartment is full of people. Andrei Tverdokhlebov is fussing with the telephone; the cord has been cut. We find out everything that has happened, and we grow numb.

It seems that before our call, the apartment had been entered by people calling themselves representatives of the Palestine terrorist organization Black September. They demanded that Andrei refrain from making statements on the "October War in Israel." They threatened to kill him and his family. Then we called. The terrorists were apparently frightened. At gunpoint, they forced everyone to remain silent, and herded them into the next room. Then they dashed out of the apartment in the brief interval of time when Tanya and Grisha had left the door to come and meet me, and before those I called arrived.

A frightful story! When the police were informed of the armed blackmail, they reacted very feebly. It was only toward evening that an officer appeared. In general, neither the police nor the KGB was interested in terrorists operating on the territory of the Soviet Union. You'd think it was the most ordinary phenomenon. Were they short on personnel, perhaps? And that at a time when many people who didn't carry arms but merely *thought* (the so-called dissidents) were being followed by swarms of "dawdlers."

One might have believed in the "initiative" of the Arab terrorists if there had not been a whole series of pressure campaigns, one after another, with a single aim: to frighten Andrei. These campaigns included the newspaper uproar in August of that year; the repeated summoning of Lucy (Elena) Bonner to the KGB office to be interrogated as a witness; a note from an unheard-of "Christian Union" with an unchristian threat to kill Matvei (the grandson); the appearance in Perovo of two men threatening Efrem (the daughter's husband) with his own death and that of his child if Sakharov did not cease his activities; the institution of "criminal" proceedings against Tatyana (the daughter) and Efrem, and the driving of them to forced emigration; the expulsion of Alyosha (the son) from an institute, and his forced emigration. And, finally, the pressure (which is still being exerted) on Andrei and Lucy via the rightless position of Liza (the son's fiancée), who is not being permitted to emigrate and join her betrothed. These are only certain "measures" of a single plan. There were also official talks at the Union Procurator's Office with an unequivocal demand that Sakharov keep quiet. And now, Gorky, under house arrest and continuous surveillance. Such is the life of the Sakharov family—a life demanding constant tension, self-sacrifice, and heroism from all its members.

MARCH 1976. The blackest day in the life of our family. While on a business trip, Grisha suffered a cerebral hemorrhage and died in a Saratov hospital. What had happened could not be believed or understood.

Throughout the frightful day of the funeral service and the cremation, two Andreis held my hands: my nephew Andrei, and Andrei Sakharov. Even after everything that had to be gone through was over, I was left with a feeling of the unreality of what had happened. At home, life went on as if the master of the house had not yet returned from his trip. There were even more people than when he was there. These were friends coming with their heartfelt consolations. Grief brought me even closer to those who had loved Grisha, including the Sakharov family.

Five years have passed since that catastrophe. Everything we have lived through, we have lived through together. And I never have any question as to whom I should take my doubts, my anxieties, my fears.

Once, on my way to the Sakharovs, I overtook three people who struck me as unusual: a young, tall man; a very beautiful woman in a

hat, also tall; and an elderly woman. They had stopped at a traffic
light on Obukhov Street, waiting for it to turn green. They were in
no hurry. Rushing past them were worried, tired-looking people who
paid no attention to the traffic light nor, apparently, to them. I too
rushed past. A short time later, when I was already in the Sakharovs'
apartment, the doorbell rang, and in came the group I had found so
unusual. The beautiful young woman turned out to be Joan Baez, the
singer.

As it emerged later, she not only sings, she engages in social action. She wanted to get approval of her position. She had expended
a lot of effort trying to convince the American government to disarm
at any price, setting an "example for the other party." She could not
believe that such a good impulse would not be accepted and would
not lead to universal disarmament. Andrei patiently explained his position to her. The circumstances under which this took place were
determined by the investigation of the Ginzburg and Orlov cases.
The very difficult life of the Sakharovs also said something to her.
She was terribly sad. She wept, and nothing could console her. Then
she sang, and her voice captivated everyone.

MARCH 1980. Andrei was already in Gorky. I was heartsick, and
very much wanted to see him. On one of her trips to Moscow, Lucy,
unable to bear up under my interrogative glance, said: "All right . . .
we'll go. Let's try to do it quietly."

At the station platform we behave like strangers. Fear nibbles at
us: What if they should suddenly take us off the train? . . . There
are three of us going: Lucy, Liza, and I. (At that time they still let
Liza into Gorky; and like Lucy and Ruf Grigoryevna, she lived part
of the time there, and part of the time in Moscow.)

Once on the train we relax. We doze and chat with a woman
sitting next to us. We go through Petushki, Vladimir, Kameshkovo,
Kovrov. Toward evening we reach Gorky. Andrei meets us on the
platform, and busies himself with the luggage. We have brought food.
The luggage has to be hauled through an underground passageway.
Andrei has a very weak heart, but as always he takes the heaviest
thing. We scold him for that.

As soon as we reach the square near the station, where we bargain with a taxi driver, we are surrounded by a herd of "the boys." From their categorical tone, we immediately realize they are KGB agents. They demand that Andrei go to Shcherbinki (the part of Gorky where he now lives), and that I return to Moscow. Andrei is enraged, almost stamping his feet on the ground. He shouts that in that case he'll go to Moscow with me. I try to calm him down. In the midst of the confusion, they look over my passport. Lucy explains. "She's his sister," she says. "Yes, his sister," I confirm. For what am I, if life has not made me his sister? The agent in charge is convinced not by our words but by Andrei's decisive tone. After vacillating, he permits me to go to Shcherbinki, with the understanding that tomorrow evening I will return to Moscow. We go off in the taxi almost happy. We talk about trifles. We make plans to buy me some shoes the next day.

The nocturnal city flashes past the windows. Here is Gagarin Street, where Sakharov's prison/house is located. We get out of the taxi, and fall into the hands of the police. Captain Snezhnitsky demands that I go to the police sentry post. Naturally, all the others follow me. Andrei takes me by the arm. "I could take you by the arm, too," says the captain. "I'm a man, too." Throughout the rest of the conversation, his tone is insulting. He becomes, in fact, so persistently insulting and crude that I cease reacting to him. Obviously, he is permitted to take such a tone. Perhaps he is programmed for it? Andrei and I exchange a few human words, and are glad that we can see each other even under these conditions.

After a certain length of time, perhaps an hour, during which the KGB boys run here and there and are plainly consulting over the telephone, the man in charge makes a decision: I must go back to Moscow today, under their supervision. During that time Lucy is throwing together a kind of dinner for me, and Liza and Andrei are giving me tea to drink—at the police post, since I have not been allowed to enter the apartment.

Embarrassed, we say goodbye, and I get into the car, with KGB agents on either side of me. In the front sits Captain Snezhnitsky. Once again the tires squish through the nocturnal streets of Gorky. I'm being taken to the station.

Later, I tried to get permission to visit Andrei from Andropov of the KGB and Rekunkov of the Union Procurator's Office. I went here and there to reception rooms, and wrote out requests. In the waiting room of the KGB an official named Andrei Anatoliyevich Ivanov told me that I could not see Sakharov in Gorky because I was an "antisocial element."

From the heap of memories of the recent past, I have chosen very little. Someday, when life is easier and I have some leisure, I will write in detail about the feelings experienced in our difficult times, and on that basis historians and writers will reconstruct not only the succession of events but the taste of our bitter age.

For that matter, all past generations have considered their own times no better. To us, looking at our age from inside, it gives no respite, flinging us from despair to hope. I am an optimist; with me, hope prevails. Not that I am convinced of a "radiant tomorrow." But the historical "today," for all its incomprehensible and absurd harshness, is different from the total hell of the evil Stalin era. As an optimist, I want to believe that in our age the rising evolution of a socialist society, too, is possible.

The fact that such a man as Andrei Sakharov—a scientist and thinker—should have thrown all his being into the struggle for human rights, for me represents a radiant banner of the times.

And the fact that his efforts are valued by his contemporaries, that Sakharov the defender of rights is known to the whole world, is also a banner of the times.

EVGENY GNEDIN

Andrei Dmitriyevich Sakharov in Exile

Andrei Dmitriyevich Sakharov is in exile. Under strict surveillance. Deprived of correspondence and contact with people in general, not just with scientists and friends. Cut off from the world. And yet one can with full justice apply to him what Anna Akhmatova said about Leo Tolstoy: "Of course he can always be heard and seen—from any point on the globe—but as a phenomenon of nature; well, like winter, autumn, the dawn."[1]

The similarity between our great contemporary and the great Russian writer of the past lies in the tremendous social significance of their moral makeup and achievements, and in the fact that both of these spokesmen for the hopes of mankind were in conflict with the regime of their time. This tragedy has a universal meaning.

People convinced of the immutability of those values whose humane and courageous champion Andrei Dmitriyevich Sakharov has become are now heart and soul with him. They send him their thoughts on his sixtieth birthday, and wish him strength and health.

[1]Lidia Chukovskaya, *Notes on Anna Akhmatova*, II, p. 30 (YMCA Press, 1980).

LARISA BOGORAZ

From "Thoughts on Progress" to "The Human Rights Movement"

"Thoughts on Progress, Peaceful Coexistence, and Intellectual Freedom" was the first work of Andrei Dmitriyevich Sakharov's that I read, early in the summer of 1968. The friend who gave me the essay said at the time: "The author requests his readers to make their own comments, expressing their views on what he has written."

Needless to say, when I picked up Sakharov's article, I was expecting discoveries. And I admit that on first reading I was not able to evaluate its worth. First of all, I was amazed by the title itself: "Thoughts . . ." For me and many others, 1968 was a time of action. Arrests, searches, trials, camp problems—this area of Soviet life, which up to then had been a "forbidden zone," was opened up for general review. The authorities' efforts were aimed at stretching barbed wire all around the area once again, while our efforts were aimed at preventing the closing of the breach. The Prague Spring and the danger that Czechoslovakia would be occupied had sparked new emotions and also prompted us to action.

I still consider this period very important in the life of our society: it was the beginning of our nonviolent resistance, of our open, free speech.

One shouldn't assume that the participants in that resistance acted only under the influence of emotions—there were, of course, discussions and arguments, and not only about specific events. Still, it is perhaps true that we weren't *thinking* much. Later, in the '70s, the article by L. Ventsov (Boris Shragin) "To Think!" appeared. In the meantime, protests, statements, open letters, declarations, demonstrations on Pushkin Square, information, the boom of samizdat.

Thoughts? What was the author thinking about? About "Progress, Peaceful Coexistence, and Intellectual Freedom." Wasn't that too general? To whom was the author addressing his recommendations and warnings? If it was to me and people like me, it was probably wasted effort. The necessity for the rapprochement of the two worlds "on a popular, democratic basis under the scrutiny of public opinion" was self-evident. I was already in favor of putting the Declaration of Human Rights into effect everywhere. I, too, considered that only in this way could one "prevent the aggravation of the international situation." But none of this depended on me. And as for those readers "in civilian clothes" upon whom progress and peaceful coexistence depended, such arguments didn't touch them. They had other values and criteria. At any moment they would sacrifice peaceful coexistence for the sake of their "concrete goals and local tasks." Then to what end was the author expending his ardor? A naive man, not of this world. Yes, not of this world, but from a higher one, where the criterion of truth is not "For what?—for a specific good," but "What?— the Higher Good." I understood this much later. For that matter, Sakharov himself cleared up my puzzlement. In 1973, in an interview with a Swedish radio and television correspondent, he said that when nothing can be done to improve a bad situation, one should "create ideals, even if the direct means to their realization cannot be seen. Because if there are no ideals, there is nothing to hope for."

Peaceful coexistence; the rejection of any confrontation, including ideological confrontation; international efforts in the struggle against hunger; the timely scientific solution of problems of ecology; implementing the Declaration of Human Rights; the democratization of society; raising the level of intellectual freedom—such are the ideals proposed by Sakharov in his 1968 article. Read his articles, his statements, his book *My Country and the World:* he is true to those ideals to this day.

Yes, of course, they correspond to your notions, reader, and mine. But they were raised to the rank of *ideals* when they were formulated by Andrei Dmitriyevich Sakharov. Before that they were *ideas,* and often seemed too abstract to affect each person directly. I, for example, considered ecology the domain of specialists: they themselves decide what is good for a person—for me—and what is bad. Indeed, human rights may be considered the domain of lawyers, and peaceful coexistence that of diplomats (or the military, if you will), while questions of ideological freedom are no doubt the province of the KGB. After all, many people look at those things that way. Andrei Dmitriyevich does not call upon everyone to bear his share of responsibility. Whether we will or not, whether we participate or decline, there is no getting away from responsibility. When a catastrophe looms, imminent or eventual, just try to claim: "I'm not involved." Sakharov assumes that our readiness to have an opinion on the main questions of the life of mankind and of ourselves is no less than his own.

But in addition to such readiness, we need the conditions for its realization: information and freedom. (Strictly speaking, for Andrei Dmitriyevich, information is the most important component of freedom.) Sakharov considers that unfreedom is the chief shortcoming of contemporary Soviet society. And this, it seems to me, is not only because of his characteristic striving for justice, not only because of his humanitarian principles, but also by virtue of his civic position. He knows that civil rights and liberties are the foundation of civic duties and civic responsibility.

Thirteen years have passed since the appearance of "Thoughts . . ." During those years there have been many changes in relations among countries in the world, changes in the situation within the country, in the fate of people close to Sakharov, and in the fate of Andrei Dmitriyevich himself. Has his position changed? It seems to me that today Sakharov's attitude toward the possibilities of socialism is considerably different from what it was before. The author of "Thoughts . . ." was hoping for a convergence of the two systems, for the democratization of the Soviet sociopolitical structure. It is also possible that he believed in the goodwill of the Soviet leadership. And then, just a few months after the appearance of his article in samizdat, Czechoslovakia was occupied. Peaceful coexistence was risked in the name of "local tasks."

Gradually, from year to year, the economic position of our country has worsened. Social inequality is growing because of the privileges given to the Party and government elite. Alcoholism and theft are becoming ever more widespread. Dedication to work is breaking down. The quality of medical care is dropping. Domestic politics ignores human rights in particular and rights in general. The authorities do not even take into account their own legislation, the laws of their own country. Citizens are subjected to discrimination on the basis of nationality or for one or another kind of religious conviction. The system of government has no feedback. Everything as a whole is based on a harsh policy of repression.

Such is socialism in our country; and apparently its defects were built into the structure created for its realization. Is it possible, then, to hope for the goodwill of its leaders, for its internal improvement? "A man may hope for nothing, yet nonetheless speak because he cannot remain silent," said Sakharov in the 1973 interview.

Plainly, certain new hopes of Andrei Dmitriyevich's are associated with the West and the possibility of its influencing the Soviet Union. But the events of the past two years—the invasion of Afghanistan, total repression within the country (including the unprecedented exile of Sakharov himself, and his isolation), and the new restrictions on emigration—all serve to undermine any belief in the possibility of improving the country by means of an influence exerted on the totalitarian regime by internal or external forces.

And yet hope remains. For me it is in Andrei Dmitriyevich Sakharov himself: in his profound morality, in his natural democratic principles and ideals, in his humanism, and even in his belief— changing, going through crises, at times losing support—a belief in the triumph of justice and progress, or at least in the possibility of those concepts being realized. What I have in mind is the evident and indisputable influence of Andrei Dmitriyevich on people around him, on those who have heard about him, on social development in our country and throughout the world. I know that many people— including myself—do not agree with Sakharov on everything. But much more important than private disagreements on particular questions is that moral potential which through him spreads through mankind.

SOFIA KALISTRATOVA

Lawlessness: A Lawyer's Notes

We live in a lawless country. Not only officials but the highest Party and Soviet organs violate the laws that they promulgate themselves. There are many examples of such violations. But the most striking example of open, cynical violation is the exile of that courageous and uncompromising fighter for human rights, Andrei Dmitriyevich Sakharov.

Exile is a measure of criminal punishment. The USSR Constitution, the Fundamental Principles of Criminal Legislation of the USSR and the Union Republics (the law of 1958), and all republic criminal codes specify that justice is administered only by a court, and that measures of criminal punishment can be applied only in accordance with the sentence of a court. There is no provision in existing Soviet law for any form of administrative exile.

And yet in Moscow on January 21, 1980, in plain daylight, Andrei Dmitriyevich Sakharov was seized on the street. And on that same day (with two hours allowed for getting his affairs in order) he was, with no investigation or trial, sent under guard to exile in Gorky.

In the eyes of the whole country, of the whole world, this was an open, cynical violation of the laws. This "action" was carried out by the USSR Procurator's Office and the USSR Ministry of Internal Affairs, institutions responsible for overseeing the execution of the law and preventing any violations of it. . . . In "justification" of Rekunkov, USSR Procurator General, and Shchelokov, USSR Minister of Internal Affairs, only one thing can be said: There is no doubt that they were only dummy executors of an illegal action elaborated and actually carried out by the KGB of the USSR, sanctioned by Party leaders at the highest level.

And now Academician Sakharov has been in exile for almost a year and a half. He has not received a single answer—not even a formalistic one—to his petitions and protests, to his demands that he be given the opportunity to defend himself at an open trial. The many letters, statements, and protests of human rights activists, individual scientists, and informal associations (for example, the Moscow Helsinki Group) have not been published. They have been hushed up and left without answer.

People speaking out openly in defense of Sakharov have been persecuted. (Thus in Makhach-Kala the scientific worker Vasef Meilanov was arrested and convicted under Article 70 of the RSFSR Criminal Code after he went into the street with a placard protesting Sakharov's exile.)

The USSR Academy of Sciences has not responded to Academician Sakharov's request for defense. And yet the Academy includes prominent lawyers who understand what the law is, and what tyranny and illegality are.

The fact of extrajudicial reprisal is only one way in which the law has been violated. People exiled by a decree of a court (whether justifiably or not) have rights regulated by law, as are the conditions under which they serve out their exile. Andrei Dmitriyevich, exiled without any court sentence but through the arbitrariness of the authorities, has been deprived of all his rights.

The law specifies that the maximum period of exile is five years. Sakharov has been exiled indefinitely.

The law does not restrict the right of exiles to correspondence. There are many letters that Sakharov does not receive, including the majority of those from abroad. Even letters from his children and grandchildren do not always reach him. He has been virtually deprived of the opportunity to use the long-distance telephone.

The law does not prohibit exiles from receiving guests at their place of exile. But almost no one is allowed to visit Andrei Dmitriyevich. All attempts by friends to see him in Gorky have been prevented by force.

And if someone on the street comes up to Andrei Dmitriyevich and begins to talk to him, such attempts at socializing are likewise cut off.

According to the law, exiles have the right to move about freely within the limits of the administrative region of exile. But a round-the-clock police post has been set up at the door of the apartment where Sakharov has been settled, and he can move about the city only under heavy guard of plainclothesmen.

Searches and seizures of documents (manuscripts, letters) in the homes of exiles may be made only in accordance with legally established procedure on the basis of a warrant sanctioned by the procurator. At Sakharov's home, secret searches are made in his absence. Andrei Dmitriyevich always carried with him his most important manuscripts, scientific notes, and diary jottings, hoping to preserve them from those secret searches. But then his briefcase, containing other things of value besides his manuscripts, was stolen from him when he was visiting a dental clinic. The results of many months of scientific and sociopolitical investigations and thinking were lost. . . .

According to the law, exiles have the right, when authorized by the administration, to go beyond the bounds of the region of exile. But Sakharov was not allowed to travel to Leningrad for the funeral of a close friend. Leningrad, however, is nothing when you consider that Andrei Dmitriyevich does not even have the right to help his wife carry things to the train. He is simply not permitted to do it—he is elbowed aside.

Thus Andrei Dmitriyevich is in a much worse position than people exiled by sentence of a court. He is completely isolated from the outer world—deprived of the possibility not only of maintaining scientific and ordinary human contacts but simply of socializing with people. He is not allowed to work undisturbed. The constant feeling of being tailed, of pressure, of having no rights, of the possibility of tyranny manifesting itself at any moment, leaves him no peace and keeps him in a state of nervous tension. Andrei Dmitriyevich is deprived of skilled medical aid, since he cannot go to Moscow to consult with his regular doctors at the polyclinic of the Academy of Sciences.

Where can one find words to express one's alarm for one of the purest souls of our age? To express one's pain, wrath, and indignation against the actions of the authorities who have put Andrei Dmitriyevich in the position of an "outlaw"?

In the foreword to his tale about the times of Ivan the Terrible, *Prince Serebryany*, Aleksei Konstantinovich Tolstoy wrote:

> With respect to the horrors of that time, the author's descriptions consistently pale when set beside the historical reality. Out of respect for art and the moral feelings of the reader, he has drawn a veil over these horrors and, so far as possible, has shown them at a distance. Nonetheless, he acknowledges that in reading the sources, he more than once dropped his pen in disgust. And he threw down the pen in indignation, not so much at the thought that an Ivan the Terrible could exist, as that a society could exist that looked at him without indignation.

How can our society look without indignation at the lawless reprisals against Academician Andrei Dmitriyevich Sakharov? One can understand those who, because of a lack of information—owing to the censorship and the closed nature of our society—know nothing about Sakharov and believe the slander that is spread about him. But everyone who knows Sakharov and his fate and does not raise a voice in his defense must feel that he is an involuntary accomplice of the evil, injustice, and violence done by the authorities to one of the best and purest of our contemporaries.

A Letter to the Compilers of the Festschrift for the Sixtieth Birthday of Academician A. D. Sakharov

Dear Compilers:

I am extremely grateful to you for the Festschrift that you have undertaken, and I thank you for honoring me with an invitation to contribute. With your permission, let this letter be my contribution.

Andrei Dmitriyevich Sakharov. His personality, his deeds, his fate, and memories of a few personal meetings with him—all this stirs up both admiration and pain in me. "Spokesman for the conscience of mankind": that characterization by the Nobel Committee is absolutely true. "A just man," the late Alexander Galich said of him. And it will do dear Andrei Dmitriyevich no harm if I acknowledge here that I have noticed in him certain traits of personal saintliness. Each time I left him after a visit I was deeply moved by impressions of the charm of his personality. And I don't hesitate to say that these were *religious* impressions.

Andrei Dmitriyevich does not belong to any of the Christian churches. But he is a very great representative of the single Church of people of good conscience and goodwill that takes in all of mankind. . . . Here I must briefly explain about our discussions within the modern Christian faith. One wing consists of followers of a narrow orthodoxy who are intolerant of any attempt to rethink tradition. The other wing consists of those who think that eternal Christianity is broader and freer, that it contains everything that is noble and beautiful, everything that is dear and holy in life. Of course it also embraces the feats in the life of Academician Sakharov: his personal charity, his deep compassion for all who have been injured, his fearless courage in the struggle for peace on earth, for the health of future generations, for human rights, for man's *dignity*, which Jesus Christ consecrated by His Advent. . . . *Christ is Eternal Man!* that is the decisive principle of Christianity. In that lies the only genuine *Hope* of human existence worthy of the word. It is not without reason that the word "humanity" has become holy for all, believers and nonbelievers. *Christ is Man,* and genuine Christianity is aspiration toward an ideal, beautiful *humanity*. And wherever it is manifested, we unworthy Christians see the Divine light. In the words of the Gospel, "The spirit bloweth where it listeth." The Holy Spirit is active everywhere where we feel the spiritual Beauty of genuine humanity. Then, wrote the Apostle, *"Christ is figured"* in man. . . . I hereby dedicate these holy meditations of mine to Andrei Dmitriyevich Sakharov.

Admiration and pain. His weak heart of course suffers from compassion. A long time ago, when he was still in favor with the government, Academician Sakharov vigorously protested against nuclear testing in the atmosphere, defending future generations against the consequences of fatal radiation. During that same period he gave 140,000 rubles of his own money for charity. Then began his public statements and, to use the Old Russian Church term, his constant *sorrowing* for each one, each prisoner of conscience in our country. The most amazing thing was that he found himself in the most complete isolation among his scientific colleagues. Yes, and the truth about our people must be stated. Finding themselves the prisoners of stupefying disinformation, a great many Philistines either simply don't know anything or even wrongly understand his very noble activity. Alas! One also has to mention here some personal insults on the part of certain fascist elements acting in the name of some pseudo-ortho-doxy of their own. . . . And now, for a year and a half, Academician Sakharov himself has been a prisoner—a prisoner of truth and pathos in that nightmare, that grief of unthinkably harsh exile.

With profound respect for you, dear compilers,

FATHER SERGEI ZHELUDKOV

Victory Day
Leningrad-Moscow

ANATOLY MARCHENKO

An Open Letter to Academician P. L. Kapitsa

Dear Pyotr Leonidovich:

The latest actions by Soviet authorities against Academician Andrei Dmitriyevich Sakharov have compelled me to appeal to you with this letter.

This crime is being perpetrated openly before the whole world, and the world is furious. Only in our homeland, which should above all and before everyone else have felt a sense of shame for the deed, is there no audible indignation or protest. I doubt anyone even slightly acquainted with Andrei Dmitriyevich could believe everything that the mass media are disseminating about him. So why does everyone remain silent? Can everything be explained by common human cowardice? Do we truly deserve our own government?

I understand, of course, that you cannot order around this government. But I am still convinced the active protest of such a well-known and authoritative scientist as yourself might have a positive influence on the "Sakharov case."

Since you are personally acquainted with Andrei Dmitriyevich, there is no need for me to write about his role and significance in our revitalized society, or about his irreproachable moral qualities as a man and a citizen. But although I am no lover of high-flown words, I

Translation by Elyse Topahan.

must warn you in advance that on the subject of Andrei Dmitriyevich, I do not refrain from using them. Please, Pyotr Leonidovich, do
not attribute this to the style of your correspondent, but entirely to
the depth of his respect for Andrei Dmitriyevich. Even Herzen speaks
about such individuals with lofty words: "The appearance of people
who protest social oppression and the oppression of conscience is not
new; there have been accusers and prophets in all mature civiliza-
tions. This higher purpose which embraces an individual is an excep-
tional and rare event, like genius, like beauty, like a remarkable voice."

I consider Andrei Dmitriyevich Sakharov to be a great phenomenon who transcends national boundaries. He has risen above that
destiny which awaits every person on earth. It seems to me already
impossible either to slander him or to praise him. However Andrei
Dmitriyevich's earthly end comes about, no one is any longer in a
position to erase him from history, which he entered as a great son
of his people. This is my opinion of Andrei Dmitriyevich; this understanding of him does not allow me to comment immediately on what
recently happened. To intercede in his behalf means to associate oneself with a great man and be a participant in history.

But why do you remain silent—a worthy and respected scientist
with an international reputation? Even if the vile slander cast on Andrei Dmitriyevich himself does not enrage you, can it be possible that
you, a Nobel laureate, do not find insulting the interpretation of the
Nobel Prize as a reward for anti-Soviet agitation?

For someone following the case of Sakharov, there is another natural
question: "Why do the most prominent Western scientists stand up for a
Soviet academician while not a single protest comes from the ranks of
Soviet academicians?" Is it possible that all Western science is in the service of the CIA? Or does it consist of bewildered people who cannot rise
above the intellectual level of "students of the Bauman College?"[1]

You don't have to be a historian to comprehend that the Soviet
state has never regarded its subjects as full-fledged rational beings.
This applies equally to street-cleaners and to internationally recognized scholars. This state is all-powerful and permits itself to do any-

---

[1]The Bauman College is a high-level engineering institute in Moscow. A letter purportedly signed by some of its students denounced Sakharov after his exile to Gorky.—
Trans.

thing to its subjects. Nothing is unprecedented or particularly unexpected about the actions of the authorities against Andrei Sakharov. Except that this time, a Nobel laureate became a victim. But this just confirms what I said earlier, that it is possible to "reeducate" or "liquidate" even Nobel laureates.

Some familiarity with your biography and scientific career precludes the necessity of my dwelling at length on the question of our still recent, sinister past. But I will address certain issues briefly.

Why have so many whose names today constitute our scientific, technological, and cultural pride been annihilated by the very state they served? Not only is the state itself guilty of the violence, but so are its subjects and its victims. Each person trembled only for his own skin. Just a few possessed the courage to stand up for the doomed.

It would be shameful to repeat this history.

So all the more, I don't understand: in those aforementioned sinister times, you saved the physicist Lev Landau from the camps and possible death. You were hardly certain at the time of a happy outcome for Landau or for yourself. But that didn't stop you then. And you risked nothing less than your head. And not just your own, but also the lives and fortunes of all who were close to you. I can't imagine a worse position to be in.

But today . . . I know people who nobly survived Kolyma, Vorkuta, and other similar places in the 1930s, '40s, and '50s, but who have not managed to maintain their dignity after release in our by now almost liberal time.

Of the well-known names, it is sufficient to recall V. Shalamov:[2] he not only lived in a dignified manner—and fortunately, survived— in Kolyma, but he also created an immortal monument to its victims, *Kolyma Tales*. Then, in the 1970s, he renounced them: "The theme of *Kolyma Tales* is no longer a problem." He betrayed himself, be-

[2]Varlam Shalamov (1917–1982) was arrested in 1937 for having declared the Nobel laureate Ivan Bunin a classic author of Russian literature. Shalamov was released in 1954. He wrote numerous short stories about the camps which have appeared in English in two collections, *Kolyma Tales* (Norton, 1980) and *Graphite* (Norton, 1981). Aleksandr Solzhenitsyn once asked him to be co-author of *The Gulag Archipelago*, but by then Shalamov was too old and too sick to accept. In the 1970s, he signed a short statement renouncing his work which, presumably, averted further reprisals.—Trans.

trayed his life's work, betrayed hundreds—no, thousands—of martyrs. . . . For what? I cannot understand. They say he was enticed
by the publication of a collection of his poems.

In the 1930s, the geologist Bratsev worked in Vorkuta as a
civilian, but all of his colleagues were prisoners. He behaved
completely honestly, and, by the standards of the day, coura-
geously: he passed along letters, he brought provisions to fellow
workers. But a few years ago, he gave a speech, in some tele-
vised jubilee celebration, about how enthusiastic Komsomol
members had conquered the Vorkuta tundra. He could have refused to participate. Nothing would have happened to him. But
if he had refused, the "Vorkuta jubilee" (can you imagine such
an expression?) would have been celebrated without Bratsev. Could
he have allowed that to happen?

Forgive me, Pyotr Leonidovich, but I must add your name to the
ranks of Shalamov and Bratsev, and what is more, of Blokhin![3]

There was a film about you on TV not long ago. A good film. You
were noble and cultured. You made quite a few wise and witty remarks. And there was even an allusion—though it was obvious only
to an informed person—to a difficult period in your life which, thank
God, passed without leaving a trace (or so the film implies). It would
seem one should be pleased when television glorifies and popularizes
a man like you, and not Ovchinnikov or Fyodorov.[4] And yet I was
ashamed of you.

The film, dedicated to Nobel laureate Pyotr Kapitsa, was presented at the very time when Nobel laureate Andrei Sakharov was
being seized and thrown out of his home, out of Moscow, out of the
institute. They filmed you, I understand, at another time; but they
still let you know about the broadcast.

---

[3]Nikolai Blokhin is the president of the Soviet Academy of Medicine.
[4]Marchenko is referring to scientists who have advanced in their careers as a result of
their usefulness to the Party and not as a reward for genuine scientific achievement.
Blokhin has already been mentioned. Yuri Ovchinnikov is a chemist and serves as vice-
president of the Soviet Academy of Sciences. Evgeny Fyodorov is a geophysicist who
has served as the director of important meteorological research centers. In 1957, he
headed Soviet participation in the International Geophysical Year. Specific reasons for
Marchenko's denunciation are not known.—Trans.

You understand, Pyotr Leonidovich, that one gang is operating here: some try to put a gag on Sakharov's mouth, while others put on a wonderful façade with an accomplished portrait of Kapitsa. And then a third academician, Blokhin, declares grandly: "The whole world knows that the Soviet state not only proclaims but also guarantees the most complete and actual complex of rights for the citizen of the socialist society." You don't believe it? Take a look! Pyotr Leonidovich is held in such esteem!

I do not know of your personal relationship with Andrei Sakharov. I do know that you once refused to take part in the development of thermonuclear weapons, while Sakharov became if not the "father of the Soviet hydrogen bomb," then one of its leading creators. You suffered for your principles then while Sakharov made a brilliant scientific career for himself. I am on your side in this matter, although my motives may not coincide with yours. Sakharov not only did not intercede for you then, but took no notice of your expulsion from science.

One could say that now you and Sakharov are even.

But think, Pyotr Leonidovich—Andrei Dmitriyevich still had a lot of time ahead of him in which to become the Sakharov of today. Unfortunately, you don't have much time ahead of you. Now is the time to think of your soul.

If the authorities were to treat you unjustly today, would Andrei Sakharov remain indifferent? I am convinced you yourself do not doubt the answer to this question.

Of course, you can reassure yourself by saying that unlike other academicians, "I have not signed anything against Sakharov and I will not sign anything." Indeed, many academicians in Sakharov's case proved themselves to be scoundrels.[5] But you are not in their league, Pyotr Leonidovich—much more is asked of a good man. Let others try to be like you. You should set a worthy example for your colleagues, young scholars, and students.

Look what they write—"students at the Bauman": "We . . . ask

---

[5]Several members of the Soviet Academy of Sciences, including Mark Mitin, a Marxist philosopher, made radio broadcasts denouncing Sakharov. —Trans.

once more to show how the West lies, defending slanderers and renegades."

Well, it is clear such people found models in the Academy other than yourself. But the fate of other young people, including your pupils, depends on your current position. The situation is pushing them to extremism; I have already heard from young people that nothing else remains for the country but to throw bombs: the violence and cruelty of the regime can only be opposed with the same methods. Reprisals against Sakharov and other members of the moral opposition to the regime—while such respected people as you refuse to intervene—will ultimately lead to terrorism, and then everything will repeat itself from the beginning. And the new Kibalchich[6] will make a choice between science and pyrotechnics in favor of the latter.

I am absolutely convinced, Pyotr Leonidovich, that your active intervention in the "Sakharov case" would really change the situation, would influence for the better the fate of Andrei Dmitriyevich and, at the same time, the fate of Russia's social development.

It is not for me to decide what form such active intervention might take. I am not an academician, not a scientist, not a laureate. "Criminal recidivist"—so the Soviet News Agency has labeled me. (Don't argue: five convictions, the sixth is already promised.) I will not lose this title under any circumstances, whatever I write, whatever I say. The only thing that I have right now is this goddam freedom which I risk exchanging for some goddam prison. The choice is mine—I do not decide for others.

However, there was a time when Russian academicians left the Academy, professors quit the university. But that was a different Academy, a different intelligentsia. Is it possible that the Soviet Academy will go down in history only for its active or passive participation in the destruction of the nation's finest sons?

They took part when Academician Nikolai Vavilov starved to death

[6]Nikolai Kibalchich (1853–81) was a talented chemist who joined the People's Will, an underground group that turned to violent methods after legitimate efforts for reform were blocked. Kibalchich set up secret laboratories to produce explosives. He was hanged after the assassination of Alexander II.—Trans.

in Saratov prison, a scientist who gave all his talent in the fight against the world's hunger.[7]

They took part when Academician Pyotr Kapitsa was thrown out of science, thrown out of the institute which he had founded.

They take part when the foul wretch gags the mouth (and binds the hands) of Academician Sakharov, with whose voice a numbed Russia had begun to speak.

Well, isn't the wise Vladimir Ilich Lenin correct: "The intelligentsia is not the brain of the nation, but the shit!"

Fortunately, he was wrong: our intelligentsia had Pryanishnikov and Kapitsa, it has Sakharov, Orlov, Kovalev.[8] But perhaps it is not yet time to reckon Kapitsa in the past tense, Pyotr Leonidovich? Gold, as the saying goes, even shines through shit.

Knowing that you are busy, I do not expect an answer, and the letter is not personal.

Respectfully,

ANATOLY MARCHENKO

March 1, 1980

[7]Nikolai Vavilov was the leading Soviet geneticist of his time. He was arrested in 1939 and died in prison in 1943.

[8]Dmitri Pryanishnikov (1865–1948) was a prominent Soviet biologist, a member of the Soviet Academy of Sciences, who tried to defend his colleague Nikolai Vavilov after the latter was arrested. Yuri Orlov was a corresponding member of the Armenian Academy of Science. An eminent high-energy physicist, he founded the Moscow Helsinki Watch Group in May 1976. Arrested in February 1977, he was convicted in May 1978 of "anti-Soviet agitation and propaganda" and sentenced to seven years of labor camp and five years of internal exile. Sergei Kovalev is a distinguished biologist whose work on the electrophysiology of muscles and the control of the heartbeat earned him an international reputation. In May 1974, Kovalev, together with Tatyana Velikanova and Tatyana Khodorovich, held a press conference in Andrei Sakharov's Moscow apartment, where they handed Western journalists three issues of the *Chronicle of Current Events*. Kovalev was arrested in December 1974 and a year later was convicted of "anti-Soviet agitation and propaganda." He was sentenced to seven years of labor camp and three years of internal exile.—Trans.

VLADIMIR VOINOVICH

Andrei Dmitriyevich Sakharov

I "declassified" Sakharov before the Soviet authorities did, and this is how it happened.

In 1964 (I believe it was) I was sitting in the editorial offices of a Moscow magazine; and while waiting for an editor who had gone off somewhere, I leafed through a reference book on the USSR Academy of Sciences that was lying on his desk. All members of the Academy (and perhaps associate members) were listed in that book, which gave the last name, first name and patronymic, position, address and telephone number, both home and office. I remember being surprised when I learned that Academician Sholokhov had two addresses, one at the village of Veshenskaya and the other in Moscow, the latter not listed, for example, in the Writers Union handbook. Strictly out of curiosity, I began to look for names known to me; and I suddenly noticed that not all the academicians had their addresses and telephone numbers listed in the reference book. For example, following the name Mikulin there was no address and no telephone number, only the mysterious letters OTN. That was all. Since I had once served in the air force and knew that Mikulin was a well-known aircraft designer, I figured that probably he was so highly classified because he had something to do with rocket engines, and that consequently the most highly classified academicians were those for whom no address or telephone number was given. By way of checking I found Korolev[1] (everybody knew that he was the most highly classified), and his name was followed by those same three mysterious letters. Aha! I said to myself. Now we'll count the most highly classified ones. (Apparently I missed my calling as a pretty good intelligence agent.) I began to leaf further through the handbook and came across a name unknown to me: "Sakharov, Andrei Dmitriyevich—OYaF." OYaF struck me as an even more mysterious abbreviation than OTN—perhaps because Sakharov himself struck me as more mysterious than the others.

Therefore, when I met up with a physicist I knew, I asked him who Sakharov was. The physicist told me that Sakharov had invented the hydrogen bomb; that he was a genius; and that, like all geniuses, he was a bit eccentric. For example, he went to the store for milk himself. That is, not entirely by himself, because he was always accompanied by several "secretaries" (special jargon for bodyguards) who kept their hands in their pockets, and in those hands clutched pistols with the safety catch off. It would have been easier on those "secretaries" if they had gone for the milk themselves. But why shouldn't a genius who created the hydrogen bomb behave a bit eccentrically? That is, within the limits allowed by special instructions. (I hasten to make the reservation that I did not believe that physicist; and I do not vouch for the reliability of the information he gave me.)

In 1968 Sakharov's name became known to the whole world after the appearance of his article "Thoughts on Progress, Peaceful Coexistence, and Intellectual Freedom." He aroused the curiosity of many people, including myself.

Five more years passed by, and Sakharov had already become a quite legendary figure. Some of my acquaintances knew him personally. I had not had occasion to meet him; and to go to make his acquaintance specially, so as to "express admiration" or "shake hands," was something I couldn't do. (Nor do I like it when someone comes to me with that aim.) But I had constantly kept track of the social

[1]Sergei Pavlovich Korolev (1907–66), Soviet scientist and space rocket systems designer, designed and directed the development of many ballistic missiles, geophysics rockets, and launch vehicles, among them the Vostok and Voskod manned spacecraft launches.—Trans.

cause that Sakharov has espoused, and I had thought much about him as an individual.

Once at the Taganka Theater they were giving a premiere of something. And, as always at premieres at that theater, there were present a great many of what are called in English "very important people," including a member of the Politburo, Comrade Polyansky.[2] Among the fairly important people was my friend the well-known writer A. (I have deliberately chosen the first letter of the alphabet so that curious people will not torture themselves with vain guesses.) He was standing next to a tall man; and when I came up to him, he said: "Let me introduce you to each other." The tall man and I shook hands. I mumbled my name, and he mumbled his, which I didn't catch. I said a few words about the show, and then walked off.

The performance had taken place in the morning. After that I had some business to attend to, and in the evening I had guests. It was only when I went to bed that I remembered the theater, the people I had met there, the writer A, and the man he had been talking to. There was something about the latter that was strange. In some way he was different from all the others (including Comrade Polyansky). There was something about him . . . But that, I suddenly realized, was Sakharov!

But how did I guess it? I knew, of course, that A was acquainted with Sakharov; but then he knew almost everybody. And Sakharov had said nothing special to me. He had not expressed any great ideas. He had merely mumbled his name, which I didn't catch. Then why did I now realize that it was he?

I will explain: *Because he bore the imprint of a great personality.*

I have had occasion, in my life, to meet several outstanding persons. And I make bold to affirm that among them there was not a single one with an ordinary and hypocritical face. Ordinary and hypocritical faces are found only on ordinary and hypocritical people.

The next day I called A to check on my guess. "Why did you do that?" he asked me reproachfully. "You immediately turned away and walked off. Andrei Dmitriyevich was quite astounded."

---

[2]Dmitrii Polyansky (1917–), deputy chairman of the USSR Council of Ministers and member of the Politburo 1962–76.—Trans.

I felt frightfully awkward. Sakharov's situation was already such that many people were afraid to socialize with him. He must have thought that I was, too. . . .

To put it briefly, I took advantage of the first pretext, called, and began from time to time to visit the famous apartment on Chkalov Street.

I can't say that I became a friend of Sakharov's; and I'm not even convinced that my visits were necessary to him. But all of my new little books that were published abroad (since there were few of them) I first brought to him. Sakharov gave one of them to somebody to read. It was seized during a search at that reader's home, and now— inscribed by me as a gift—is in the files of the KGB.

Earlier I said that in the course of my life I have had occasion to meet several outstanding people. But I have known an even greater number of famous people—sometimes famous throughout the world. I hope it's clear that "famous" and "outstanding" are not always the same thing. I have known outstanding people who were known only to a small circle of acquaintances. And I have known famous people who became so by accident or thanks to their own special capacity for taking advantage of historical or other circumstances and were not ashamed (in the words of Pasternak) "while counting for nothing, to be a parable on people's lips."

Sakharov has made no special efforts to achieve fame. I don't even know who can be compared to him in attempts to minimize his own merits. From his Kremlin tribunal, Academician Aleksandrov says that Sakharov's achievements are exaggerated; and Sakharov himself says they are exaggerated. The Soviet propagandists claim that Sakharov is doing nothing interesting in science; and Sakharov says that, in general, one should work at physics while under the age of thirty. Since he is older than that, you can take his meaning as you will.

And yet one well-known physicist has told me that still today, all the main experiments with a controlled thermonuclear reaction are based on Sakharov's ideas. And it is said that even the academic higher-ups cannot but acknowledge that during recent years, in a crowded apartment, with daily crowds of petitioners, dissidents, and correspondents, while busy with his chief struggle, and constantly harassed, he regularly produced new papers with new ideas. Personally, I can't imagine how he managed to do that.

I have heard people express the opinion that human rights, about which Sakharov talks so much, are a secondary matter; that a national or religious rebirth is much more important. But without human rights there can be no rebirth. Without them there can only be either decay or, in the best (more accurately, in the worst) case, a change of ideology and the rapid movement of the masses from one swamp into another.

It is often said of Sakharov that he is courageous. But that characterization, if it does not contain a moral evaluation, is one that I do not accept. What is courage? Physical valor? Any adventure-seeker may possess that. But Sakharov does not resemble an adventure-seeker. Conscience and a clear understanding of the catastrophe threatening mankind have prompted him to follow a path where courage alone isn't enough.

The late Konstantin Bogatyrev[3] and I once went to visit Sakharov at his dacha. He met us on the platform at the train station. It was getting toward evening, and the sun, big and red, was already dipping under the horizon. "That sun," said Andrei Dmitriyevich, "reminds me of the explosion of a hydrogen bomb."

I had imagined such an explosion differently—as a fiery, boiling element. But I was only imagining, whereas he saw. And he conveyed his notion to me. Now a red setting sun always fills me with a feeling of alarm: I see it indifferently hanging above our lifeless planet.

They have expelled Sakharov from Moscow, and sealed his mouth. That is not only cruel to him, but senseless. Wherever he may be, the problems identified by Sakharov (but posed by history rather than by him) will not disappear. And the longer the people holding the fate of mankind in their hands avoid solving them, the more unswervingly will we slide down into the abyss into which he has already peered.

[3]Konstantin Bogatyrev, Moscow dissident, poet, and German translator, died under mysterious circumstances in 1976.—Trans.

We Happened to Get This Letter . . .

We happened to get this letter, most unusual both for its content and for the way it was written. It was so simple and so full of misery and a kind of lonely desperation that without trying I have remembered it with the spelling just as it was.

The letter was addressed to A. D. Sakharov, and read as follows:

To the Ministry of Rights in Defense of Man
Moscow
From Citizen Ivan Maksimovich Gushchin,
   born 1915, living in the village of Sloboda . . .

Statement

Please help look over my statement and help me about what is said below. I am a citizen of the Sovietsk Union a participant in the Patriotic War I have awards and please help get either a pension a year has already gone by and I can't get a pension I have a bad wound in the chest I'm still carrying fragments of a shell as certificates of a wound And I lack two years seniority and because of that can't get a pension because after the war I got into prison and spent 17 years there so please help me and send a explanation to the lower organs and to me too so I can some way exist today because I can't work physically after the wound and in general because of old age And in summertime

I herd cows I don't earn much and in wintertime I don't work at anything. And since I came out of prison I don't have a house or a home and during the war I lost not just my health but my family and I go on wandering like a wolf, where I have to and how I have to So my life is passing.

Please don't refuse my request

3.7.76   Petitioner I. M. Gushchin

Right now as a shepherd I sit on a stump and write.

The fact that in the thoughts and heart of that simple man, a cowherd, there dwells an awareness that there is Truth in our country, a belief that there is a "Ministry of Rights in Defense of Man," and the hope that it will help—is that not one of life's rewards?

So meaningful for all of us is Andrei Dmitriyevich Sakharov that it was impossible for the cowherd of the village of Sloboda, Ivan Maksimovich, who had heard of Sakharov as a great intercessor for people, to imagine him otherwise than as the Minister of Rights in the Defense of Man.

Sakharov has been exiled to Gorky, and a vigorous campaign is being waged against him using all the means of official mass propaganda; but Sakharov's prestige, and the respect for him, have not lessened. One can frighten people, one can force them to remain silent, but to deceive them is difficult. And people are nonetheless drawn to Sakharov. They find out his address, ask about him, and send words of warmth. And the most surprising thing is that despite his being in disgrace with the state, they place hopes in him.

The sentinel at the door is powerless against Conscience, Honor, and Dignity, which in our day in our country are called Courage.

Among people who are downtrodden, who seek support and help, there lives an image of the strong and the mighty—that is Sakharov.

Among people defending their human rights, there lives an image of courage—that is Sakharov.

And nothing can be higher than that prize, the human prize of Trust.

EVGENIYA PECHURO

A Man from the Future

A man of the future is living today, here, among us. Is that not amazing? But we, in our daily cares and concerns, seem not to notice that this is a personality that does not fit within the framework of our times. For this is a man who sees the present from the future point of view.

It is not that future imagined by utopians of various times who assumed that they knew what man did not need and, from that negative knowledge, derived normative affirmations as to what he needed. And it is not that finite "radiant future" once promised to us by the fathers of the October Revolution. It is an infinite future, the essence of which lies within oneself, in one's capacity daily to go beyond one's own limits, rejecting those forms of unfreedom of man and society that mankind has created along its way. There is good reason why Andrei Dmitriyevich's favorite idea is overcoming the divisiveness of mankind as a decisive factor on the way to peace and freedom. The indivisibility of the world, like the inseparability of its preservation from the preservation of human rights, is not just a credo preached by Sakharov: it is the object of his constant concern, the content of his life.

If the integrity of the world and the integrity of man, with all his requirements and rights, are so interdependent, how can one live by that, knowing, as Andrei Dmitriyevich has said, that "in my opinion,

almost nothing can be done"? That "the fact that we act does not
mean that we hope for anything"? How can one constantly appeal to
the authorities with different kinds of specific suggestions (e.g., about
abolishing capital punishment, or about political amnesties), and even
with entire programs (e.g., a program for solving the problem of Afghanistan), knowing that one will not get an answer and will not be
understood, since "they have another way of thinking"? How can we
understand the fact that, while affirming that, Andrei Dmitriyevich,
under all conditions (even now in Gorky, exiled, under house arrest,
without the right to visits or correspondence, under constant threat
to his own physical existence), steadily follows the line of behavior he
has worked out, preserving that elevated state of the spirit which is
simply impossible without a deep inner equilibrium? I can find only
one answer to that. The deep equilibrium without which he could
not lead the kind of life he is living is given to him by a premonition
of the future as the capacity for overcoming the present.

The process of overcoming the divisiveness of the human race and
man's alienation constitutes the essence of the history of mankind.
But people learn of its realization—always partial!—only *post factum,*
looking at yesterday in the light of today. Andrei Dmitriyevich is one
of those for whom "tomorrow" *is* "today."

Sakharov is not a superman. Of course he is a great scientist and
a man with a great soul. . . . But the true scope of his personality
is determined primarily by that very rare combination of potential
possibilities and their realization, that very rare harmony of words
and deeds, which also gives us the right to call him a man of the
future: a currently existing potential man. And not normatively, not
come to completion once and for all, but always becoming, overcoming himself, as the future does the present.

The Price of Recantation

This essay was inspired by totally different people, different deeds. But I take great pleasure in dedicating it to a man who has not recanted.

I am more and more convinced that if the human rights movement yields any political results it will be only in the distant future, in some entirely different era, in a completely different Russia. But the farther off that is, the more I like the human rights activists. What they are doing is hard to justify in terms of outer expediency. But in their hopeless cause there is something else besides the impossibility of political success: the possibility of remaining a human being—a step toward what Dostoyevsky called a strongly developed personality. Before our eyes arose a kind of all-inclusive Leo Tolstoy, with his "I cannot remain silent." And the more that all-inclusive man loses, the more his spirit grows. Thus there grew before our eyes Andrei Dmitriyevich Sakharov, losing his position, money, work, and outer freedom. He grew, and he remained each person's equal. No one is oppressed by his worldwide shadow.

We are living in an age of crisis of *all* movements. *All* radical means proposed for saving Russia or the West or mankind have proved to be ineffective or fatal.

The only world movement which has any prospects is the movement toward disintegration. It began with the disintegration of empires, it responded with the disintegration of the atomic nucleus, and where it will end, God only knows! Countermovements—toward creation, crystallization—achieve partial successes; but the overall processes of disintegration constantly threaten to wipe them out. In the sixty-odd years that I have lived through, the world has become more chaotic and more mad, and even the threat of universal atomic death has not given mankind a greater sense of solidarity. Europe has become more unified, at the expense of a tricontinental hatred of Europe. The Jews have become more unified, at the expense of Judophobia. The solidarity of the have-nots was fostered by hatred toward the haves, and is constantly nurtured by hatred of the imperialists, the kulaks, the apostates, the aliens, the rotten intelligentsia, the rootless cosmopolites, etc. In this cloud of hatred it is impossible to breathe; and unless a miracle happens, all our civilization will choke, lose its reason, and destroy itself.

What remains? To be a human being. Let Providence be concerned with saving what can be saved; our job is to remain people, drawing up our plans and carrying out those plans in a human way, without slavish fear or base compliance, without the maniacal possession of the fanatic. Only then will something, perhaps, be saved. Only people who have found support within themselves will someday form a new society without prophets and false prophets. After the turning point which cannot be held back or stopped (too many leaders are calling for it, and too many madmen are following the leaders). After the whole world's emergence from the era of sickness, which in our country has taken on its own special—almost the worst—form. For we have not digested the ideas of our radicals. We have choked on them, and we can neither swallow them nor spit them out.

I have never been very keen on the plans proposed by Andrei Dmitriyevich Sakharov, but I have always relished the man showing through the text of the plan: serene, sober-minded, benevolent, without any fears, without any hatred. What he has proposed has touched upon only the most alarming symptoms of the sickness and has not affected the deepest spiritual causes. But he has offered his simple medicine with a kind hand; and it was pure medicine, without any contradictions. What he has done has not always come out well.

Professional politicians have criticized Sakharov: he didn't defend the right people; he didn't speak out for the right thing. But for me the charm of Andrei Dmitriyevich Sakharov lies precisely in the fact that he has defended each and every one, that he has not become a politician, that he belongs to no party, and that in a world overflowing with hatred he has paid hatred no tribute. From time to time he has remained alone and absurd in his humanity, like Don Quixote.

I have never thought, and I do not think now, that the human rights movement is the only form of a worthy human life. The main thing is to have spiritual independence. How it is manifested is a matter of free decision: in protest, in politically silent but independent communes quietly doing their work, and in the solitary creativity of the writer. But if you have come out as a defender of the bound and the dumb, remember that your bankruptcy may ruin thousands! Do not deceive those who have trusted you, who have invested their treasures in your name! In the past ten years we have often seen how soap bubbles burst. How vanity and brazenness (giving themselves out to be courage and a feat) have ended up in shameful recantation. How people did not sustain their roles, their poses. How they shrank before the mere specter of violence. Andrei Dmitriyevich has never talked a lot about himself, never tried to hog the stage, never raged against weakness, never demanded firmness of others. His own firmness is all the more to be prized because of that.

In our frantic age, peoples, having lost their ancestral models, have not succeeded in acquiring another personality structure. They have become a faceless mass swinging pendulumlike from pettiness to caddishness, from caddishness to pettiness. We cannot stop that process. Only Providence can stop it. But we can see things as they are, and dash away from the sewer. We can bear up under the test of a time which has lost its dogma and find our own personal approach to the eternal spiritual sources—our own support for honor and conscience. Laws govern only mass movements. The individual person is free, and can find in himself the basis for intelligent and moral action: can become a support in a world without supports, just as Andrei Dmitriyevich Sakharov has become our support.

1. A Dialogue with the Inquisition

It will soon be ten years since the Holy See abrogated the condemnation of Galileo Galilei. Today no one remembers the text of his recantation. What did it say? Probably the same things that people say today: that finding himself deeply in error, he discredited the church . . . and it banned the use of the books he had written. . . . Those clumsy formulas are quickly effaced from memory. Only one thing is remembered: "And yet the world turns!"

Likewise, of the legal cobweb in which Giordano Bruno was caught and burned, only the words of the convicted man remain: "You are more afraid, in handing down this conviction, than I am in hearing it."

Apparently, Giordano Bruno actually said that aloud. But Galileo only muttered "And yet the world turns!" Or he thought it. Or someone else thought it and said it. But whether it was said or not, it has become a proverb. Every schoolchild knows it. Giordano Bruno's words are much less known. Why? Perhaps because modern man is rather frightened by Bruno's fearlessness. Like Panurge, we are all ready to stand for the truth, except for the stake. We're afraid of the stake. And perhaps we are even more afraid of the way to the stake—alone, subjected to the hooting of the crowd. No, such a death is never beautiful. It is hard to repeat Bruno's words. It is a bit frightening, even in thought, to put oneself in his place.

Galileo is more understandable. We too would have broken down if we had been shown the instrument of torture. But then, coming back to ourselves, we would no doubt have muttered: "And yet 2 + 2 = 4."

Because actually 2 + 2 = 4. The fact that the world turns, that the economic system is collapsing, and other such external truths cannot be proved by the fortitude of the scientist and cannot be refuted by showing the whole world his weakness. Yes, Galileo was weak; but he aimed his telescope into the depths of the starry heavens, and everyone could look and see what was going on there. Galileo's method proved stronger than the medieval method of proving the truth. And the fact that by that medieval method Galileo was forced to recant proved mortal not for Galileo but for the methods of the Inquisition, their *reductio ad absurdum*. "And yet the world turns!" rings out like the sentence passed by modern times on the thought of the Middle Ages: *"Absurdum est!"*

Galileo talked the way our relatives and neighbors talk. Martyrdom isn't necessary! The conclusions of science are enough! Sooner or later science will have its way. But Bruno silently condemns us. He was a man who defended truth in the medieval way: he gave all. He was a monk, a mystic. And his fortitude is difficult to understand without being accustomed to discipline, contemplation, and prayer, without the firm canon of behavior left by the martyrs.

For monks at each step there is rank and etiquette. Strict discipline counts more than prayer. The vows. The oath. In difficult circumstances this gives a man much support; it counterpoises his weakness. The first soldiers, who received their pay for their warriors' work, had no motherland that they loved, but they gave their lives. For what? For a little money? More likely, because of habit and discipline. And because they were prepared for that death in battle and didn't fear it. Such a death was a condition of their profession, of their honor, as a doctor is ready for cholera, a nobleman for a duel, and a dissident for arrest, investigation, and a trial. Perhaps the soldier loved his motherland, and the dissident loved freedom (or that same motherland), but what helps him to hold firm is a steady purpose, a law of honor, an oath: never talk with them! What caused officers who had faced death at Borodino to break down?[1] The lack of a code. The lack of rules for behaving while under investigation. The attempt to find a common language with Nicholas.[2]

Galileo was not accustomed to law, regulations, etiquette. It seems to me that he was even repelled by all that. I imagine him as natural and direct. And in a difficult situation, the most direct thing is weakness. Bruno was a monk. Campanella was a monk. And both bore up. Bruno may have had special sources of moral inspiration, but in Cam-

[1] An apparent reference to the Decembrists, many of whom, after having displayed great bravery in fighting against Napoleon, broke down when being interrogated after the December Revolution.—Trans.

[2] This is not a principled rejection of dialogue. But the self-respect of a prisoner does not permit him to carry on a dialogue with an investigator. A dialogue is something that occurs only between two free persons.

panella I do not see any grace, any mystical aid, or any special faith. I see merely the habit of discipline and the understanding that to bear up under torture means to preserve life and freedom. In the torture chamber the inner voice (not very deep) may easily play one false and become the advocate of apostasy. Law is more reliable. Just as discipline is more reliable in war than patriotism.

But if the inner voice is very strong, one can do without discipline. Churikova, playing the role of Joan of Arc in the film *The Beginning*, is afraid of torture, afraid of the stake. Probably the historical Joan was also afraid. But even stronger than that base fear was the noble fear of betraying her visions, of calling those angelic visions diabolical. They touched the soul very deeply, they came from a very great depth. The fact that certain present-day neophytes have recanted (without even having seen the instrument of torture) has simply shown the world that they were not in a state of grace. There were exaltation, theatrics, hysterics, and delusions of religious grandeur. There was the appearance of faith. At the moment of truth, alone with the four walls, one self-deception yielded to another. Before, they took refuge in the Apocalypse; now they concealed their apostasy with a quotation threadbare as a proverb: "There is no power except from God."

The third way to firmness is the simplest. If there is no help from God, one can grasp at the devilishness of the torture, at the bitterness of the fight. Here there is one danger: how to get through the pause— a month, two, three—without a struggle. Here the devil is not on the side of the prisoner. More likely he is helping the investigators. But the bitterness helps one to bear up under what seems to be the most frightful thing. I was told about a Soviet intelligence agent who got caught and bore up under third-degree interrogation by the Japanese. How did he do it? He constantly swore in obscenities. The agent was a Georgian, but he swore in Russian. There is a kind of sorcerer's strength in obscenities: —— your mother,—— your mother, all in succession, without shame or pangs of conscience. Fear inhibits sexual aggressiveness, and vice versa. Swearing with unbridled, unrestrained sexual aggressiveness inhibits fear, forces one to throw oneself on the danger, like a salmon up a waterfall. To smash oneself to death or break through into the reservoir and release one's fluids over the eggs. "—— your mother" is a magical phrase of devils. In cursing oneself and others with that sorcerer's phrase, one can put down rebellion. (Shchedrin wrote that the first word used by an experienced administrator in speaking to a crowd of rioters is an obscenity.) Terrified new recruits who have just fled headlong from German tanks can be stirred to go over to the attack by obscenities. (I have undergone this, and speak from experience.) In combat, obscenities make up more than 50 percent of the words used by officers. With this kind of cursing we conquered the Germans in the Great Patriotic War, and with it we overcome economic difficulties. "—— your mother" replaces material interests. (Apparently this fact has not been taken into account by a single economist. I ask that it be considered my contribution to political economy.) "—— your mother" provides the nerve for rush work at the end of a plan period, when with convulsive efforts the lacking percentages are turned out. With that same sorcerer's phrase a tractor driver who has been issued a machine according to plan—a tractor in which one-fourth of the parts do not fit each other—adjusts them, gets the tractor into shape, and plows the field of real socialism.

Why hasn't that worked in China? Perhaps there was simply a lack of linguistic resources. They began looking for something else. (Mao tried the Trotsky model; Deng is trying the Bukharin model.) And to whatever extent the Soviet (basically Stalinist) system has prevailed in the world, it has only been by the influence of Gennady Shimanov,[3] only given a certain russification of the world. One can imagine a real socialism without Marxism, but not without Russian profanity. The small Sovietized countries prove nothing. They subsist on Soviet doping, and the Soviet regime subsists on the resources of the Russian language. Obscenity is not at all something from the people opposed to anything Soviet. What is opposed is the religious verse about the Mother of God who brought out of hell all those who had never cursed in an obscene way. But obscenity (*mat*) is that which has beautifully combined with dialectical materialism (*diamat*). As was said about that a long time ago, they swear in obscenities (*mat*), but they

[3]Shimanov is a contemporary representative of what has been called "the Russian New Right." His outlook is an extreme form of Russian nationalism.—Trans.

take refuge in dialectical materialism (*diamat*). Both constitute a mighty weapon for the proletariat.

Did Giordano Bruno feel the bitterness of the fight? I think he did, and that in his breast there again flared up the fire of the dogmatic quarrels which once impelled a legendary prelate to pull out a tuft of Arius's beard. I believe that Bruno had everything: law, the grace of God, and (at certain moments) the violence of ecumenical councils.[4]

Thirty years ago I said something different. But when my cellmate (companion on the plank beds) rated Galileo above Bruno, I hotly championed the burned heretic.

Today I don't want to argue over which of the two is superior, but to understand why the martyr Bruno and the recreant Galileo stand together in our memories and not opposed to each other. I thought and thought about it, and I had a dream. "You know," someone said to me, "only a few dinars are made out of pure gold. Most are made of dark gold. But the dark dinar is worth only one dirkhem less than the pure one." When I woke up, I could not remember the monetary system of the caliphates (a dinar was worth ten rubles, and a dirkhem one ruble), but the sense of the dream was clear: a dark dinar is still a dinar, although it is worth a bit less than the pure one.

Not all truths hold up under recantation so painlessly as did the heliocentric system. Some truths simply cannot be proved or refuted, but only strengthened by the fortitude of their confessors or weakened by their weakness. Recantation is nothing for a scientific theory, but it is a great lesson for a moral truth. Socrates understood that. His cup of hemlock confirmed the freedom of moral investigation more than the entire library of Alexandria. And yet, we don't condemn Anaxagoras, who fled Athens. And we don't condemn Uriel Acosta.[5] Amid the general loathsomeness and apathy, a person who has taken upon himself the labor and risk of free thought is worthy of our sympathy, even if he has proved too weak for his task, has proved morally

---

[4]My teacher Leonid Efimovich Pinsky bore up under fifty-six nights of successive interrogations without sleep. He was helped by the polemic vent, the violence of the quarrel.

[5]Uriel Acosta, or la Costa (1585–1644), philosopher and teacher of Spinoza, was persecuted for not believing in the immortality of the soul, and committed suicide.—Trans.

and intellectually unprepared for the cause he has taken up. Even if he only imagined that he was capable of sacrifice and did not bear up under the ordeal. Even if his theory is not well perfected and has begun to fall into pieces in his shaken mind. Only recantation without pain, without sorrow, is loathsome. We feel loathing for the recreant who is complacent and grateful to his jailers because they have afforded him the time and place to engage in theology, and who preaches to others: "There is no power except from God."

Gutskov's play *Uriel Acosta* was translated into Yiddish, and I saw it at the theater where my mother acted. She herself did not act in that play, but the role of Ben Akiba (a kind of Grand Inquisitor) was played by her second husband, Oibelman. He was a very talented actor. I was amazed by the impassivity with which he said: "Everything has already been once before." There were atheists, and there were heretics. In real life, Oibelman went mad a few years later from the fear that he would be arrested; and for fifteen years he lived through a kind of eternal interrogation, hearing voices reminding him of the trivial delinquencies in his own life. (My mother took pity on him and did not put him in a madhouse, where he would not have lived long.) But on the stage he managed to create an image of the unshakable wisdom of the community, its great WE, going back deep into the millennia. And yet that WE could not counterpoise the weak, trembling I of Acosta. They tempted him with a visit by his blind mother, who begged him to recant. They tempted him with a visit by his fiancée, who enticed him with her love. Acosta repeated with her the eternal words of the Song of Songs: "Thou art fair, O my beloved, and thy lips are like honey to me. . . . Thou art fair, O my beloved, and thine eyes are like doves."

The man's heart trembled, and he repudiated himself. Then they betrayed him (they gave the girl in marriage to someone else).

Acosta remains alone. Apparently he was silent for many years. But in the last scene we see him with a pupil. And the name of that pupil is Baruch Spinoza.

Why did Spinoza bear up under threats, excommunication, and anathema? Why didn't he recant? Perhaps because he was not torn between two passions: for the truth, and for a woman. Or perhaps he knew the price of recantation even before they de-

manded that he recant, and realized that he could not pay that
price. Recantation is not a maneuver, not a tactical withdrawal in
war. It is moral death. That death may be followed by resurrection, and it may not. Only complete honesty alone with oneself,
only keeping a deep and complete silence, only the ability to
drink everything to the bottom, can save the soul; and external
defeat becomes a step toward internal growth, toward that point
of equilibrium at which a person looks at himself with the eyes
of God.

A publican can raise himself higher than a Pharisee, but only under one condition: if he recognizes, deep down, that he is a publican.
Weakness, honestly recognizing itself to be weakness; recantation,
honestly recognizing itself to be recantation—these things are not yet
mortal sins. They are bitter, but they may be healing for the soul.
Recant, and be silent. And let the bitterness of recantation mature
within yourself. Perhaps a new strength will grow from it. An inner
strength. If you have been able to pay the price of recantation to the
fullest. Very few people have managed to pay that price. But it is
measurable, and I can believe that a person can bear up under it.
And what is the price of recantation for those who have obtained it,
those who sent Acosta's mother and fiancée to him, who played on
the heretic's weakness, impulsiveness, and childish directness? I don't
know. I can't put myself in their places. I know only one thing: that
the world has forgiven Galileo his weakness, but it has not forgiven
the Inquisition its strength.

2. *There Is No Power Except from God*

Weakness is worthy of commiseration, until it begins to admire itself and tries to prove its superiority. The Pharisaism of the publican
is worse than the Pharisaism of the Pharisee. The Pharisee is at least
proud of something that in itself is good: his fidelity to the law. But
what is the publican proud of? Of his seeming similarity to his fellow of the Gospels. But the publican of the Gospels did not yet know
the parable about the publican. The Prodigal Son of the Gospels
threw himself at his father's knees without counting on the fatted
calf.

Counting on the fatted calf remains counting on the fatted calf, regardless of what word is used to conceal it. Any phrase in the mouth of a scoundrel becomes scoundrelly, even one taken directly from the mouth of God.

There are certain sayings that scoundrels especially like to quote. One of these is: "There is no power except from God." These words belong to the Apostle Paul. But what of that? Close beside them one can find other words that are diametrically opposed in meaning: "One must heed God more than people." Both sayings are found in the Acts of the Apostles, edited by the pupils of that same Apostle Paul. Such phrases are responses within a dialogue. Outside of that dialogue, out of context, they are not true and not false. In one case, "Not peace, but a sword" is true. In another, "He who lives by the sword shall perish by the sword" is true. The truth of such quotations depends upon how accurately they are applied to the matter at hand. One can take a phrase from Holy Scripture and cover it with filth. One can take a saying from the mouth of a fishwife and utter the holy truth.

Out of context, a categorical judgment is true only when it does not aim very high and connects or distinguishes between clearly defined objects. (There are cones on a pine tree, and leaves on a birch.) With the passage of time, people have learned how to define and connect objects going far beyond the framework of common sense. This has been achieved with the aid of the exact sciences. But the scientific law $E = mc^2$ is only a strictly thought-out and strictly formulated summarizing of experience as simple as the lack of leaves on a pine tree. In establishing such laws, no one quotes Holy Scripture. That which is *per se* clear or strictly demonstrable does not need the support of authority.

A reference to authority points to another class of categorical judgments which are not obvious in personal experience, cannot be demonstrated, and, to put it honestly, are not always strictly true but must be taken as true. In particular cases, guilt may be established, but the presumption of innocence does not thereby lose its force. It remains in force as a mainspring. In particular cases, jurists are entitled to say "not guilty" to an evident murderer, an evident thief. The tsar was entitled to pardon Katyusha

Maslova.[6] But the law does not thereby lose its force. In the New
Testament the right to violate the law is associated with grace. Grace
is higher than the law, but it does not replace the law. Christ healed
on Saturdays, and did not throw a stone at the woman who had sinned,
but he did not annul Saturdays. He said that he came not to violate
the law but to carry it out. Through grace, the mainspring may be
stretched, but it must remain strong, must return to its place with
great strength. And here a reference to Scripture is needed. A commandment must always be remembered. Even when you are violating it. Especially then.

But Paul's words in the Epistle to the Romans were by no means
a commandment. Rather, they were a rebuke. To the Zealots, who
dreamed of rising up and (with the aid of legions of angels) breaking
the power of Rome. And a rebuke to the Chiliasts, who dreamed of
leaping immediately into the millennial kingdom of the just. Paul was
a realist. He had in mind what was later said about historical necessity. But he by no means commanded kowtowing to the Romans.
Subordination to temporal power has its limits. (Christ said one should
render unto God what is God's, and unto Caesar what is Caesar's.
Paul said one must heed God more than people.) Civic discipline did
not mean rejecting the freedom of preaching. As that was understood
by the apostles and the martyrs, so it was understood by St. Ambro-
sius, who did not enter the temple of the Emperor Theodosius, and
by St. John Chrysostom, dismissed and exiled for exposing the imperial court.

No necessity or predestination or will of God eliminates human
freedom. The kingdom of freedom exists within certain limits, but it
always is. There is the possibility of choosing one of several terrestrial
paths (at the worst, death, like Cato). And there is the possibility of
opening the door into the depths, of finding the boundless expanse
of inner freedom. Logically, freedom and necessity, predestination
and freedom are in a difficult relationship, but that does not mean
that they exclude each other. In various traditions, in various cul-
tures, freedom and necessity wear different clothing; but they are

---

[6]The heroine of Tolstoy's novel *Resurrection*. —Trans.

always together. They are connected, like God and man in Christ, unmerged and undivided.

Translated into Chinese, "There is no power but from God" is the mandate of Heaven. The ruling dynasty has the mandate of Heaven. But if it is overthrown, there is a change in the mandate. God's assent is replaced by God's will, and there comes an end to God's will.

Just when? That is nowhere said. It has to be felt. One's conscience and one's reason are not replaced by any law, any scripture.

In Indian thought, the main thing is not what will happen with society but what you will be. The limits of personal fate are karma, that which is given. But those are precisely limits, and not the path itself. Thus we are given language. With rare exceptions, we do not choose the language of our thought. That is our karma. But in one and the same language one can stutter, use clichés glibly, or write really freely—that is, in one's own way, with talent. A good stylist is condemned to freedom. He is compelled to be himself in each phrase. A bad, bureaucratic, alien phrase tortures him like a bad deed.

The inertia of the linguistic system does not interfere with the freedom of the stylist. Likewise, the inertia of the religious system does not interfere with the freedom of the mystic, and the inertia of political freedom does not interfere with the freedom of the reformer. The will of God is not entirely gathered together in inertia. It is no less present in the violation of inertia: in Paul's resoluteness to begin preaching among the pagans, and in Muhammad's resoluteness to begin the conquest of the world just at the time when both seventh-century superpowers, Iran and Byzantium, had exhausted each other and were ready to fall at the feet of the Arabs. Paul felt not only the *impossibility* of acting like a Messiah with sword in hand, but also the *possibility* of acting in his own way—and he *acted*. Freedom is understood necessity, the understood (or felt) chink through which one can thrust and break inertia, that weak sector of the front through which one can break into the open spaces.

The parameters of freedom are changing. During the Renaissance, both the individual and the nation had more freedom than in the last centuries of the Roman Empire. But even then there were other choices besides throwing oneself on one's sword. The less the possibilities for outer freedom, the more persistent the call for inner freedom. . . .

In history there is always a place for freedom. It is only in utopia that there is no freedom. Marx's fateful mistake lay not in his study of the laws of necessity—that constituted his strength—but in what he called "the leap from the kingdom of necessity to the kingdom of freedom." "Mankind will come to communism, or will perish." There is no more multiplicity of means of development, no more choice. Or to put it another way, the choice between salvation and perishing has been predetermined. Thanks to the absolute goal, all means are good. And necessity commands freedom: Don't spare the cartridges!

Are there guarantees against that leap into utopia? Apparently not. From the chink of science grows the red specter. And from the chink of religion grows the black specter of Ayatollah Khomeini.

But somewhere along the line God's assent is replaced by God's will, the inertia of existence, which only awaits the time to run short. Real history comes into its rights, and within its necessity freedom again opens up. The individual person is now free to choose a worthy path of life and death. I do not say a sinless one, but a worthy one: in one of the humanistic professions, in simple love for one's neighbor, in protesting against evil. That is enough. The most important thing is not what will happen to us, but what we will be, whether we will be realized or not. If we have been realized, on that inner fortress we can defend even a certain outer expanse. Whatever power God has sent us. For there is no power but from God.

3. "Together with the Country, Together with All the People"

There is yet one more figleaf of Pharisaism: "I together with the country, with the people. I don't want to be a renegade, part of a fifth column."

"For goodness' sake," you ask him, "what fifth column? In what war? Who is offending you, Tit Titych? You yourself offend just anybody."

But to the Pharisee it isn't important who attacks. Under no circumstances does he want to be a renegade. He wants to be only with the country, with the people. Even with the hangmen? Yes, with them, too. If I realize that without fear of the hangman the country will collapse, it is stupid not to shake hands with the hangman. One must find a common language with the hangman. The hangman has his own sincere convictions, and one must take them into account. . . .

There is a certain logic in that. More than just a few people have lost the ability to detect falsehood. It is a mass syndrome, a kind of habitual dislocation of the popular conscience. And in our age, to live according to one's conscience means to live as a renegade.

In his essay "Walking on the Water," Boris Khazanov wrote:

A remarkable trait of our countrymen consists in the fact that they always act in accordance with circumstances. The immediate situation—that is what determines behavior as a whole, and then the way of thinking. Since that code for living corresponds to a theory, the first point of which says that existence determines consciousness, our countryman will not be offended if you explain that to him. Existence does, in the direct and literal sense, determine his consciousness. When there is room in the bus, he is a human being. When it is crowded, he turns into a beast. In general, it costs him nothing to switch from saccharine politeness to a wolf's howl. Like Proteus, he changes before our eyes, transforming himself from a humble worker into a Hun and then, when the occasion suits, just as freely taking on a human appearance. In a word, this is mass man, a kind of organism whose temperature is always equal to the temperature of the outside environment.

Nowhere does this trait show itself so clearly as in a labor camp. A camp, as it is not difficult to see, is a miniature scale model of society. However, the human material we dealt with there was not homogeneous. And it was always easy to tell a fellow countryman from a foreigner. The latter may have been cultivated city-dwellers, like the majority of those from the Baltic countries, or illiterate peasants, like the Western Ukrainians or the Belorussians; but there was always a sharp line dividing them from "ours," as though they were people from a dif-

ferent civilization. They were distinguished by the morality they
had imbibed as children. That morality, like an anchor, gave
them stability in an absurd world. And although they wav-
ered, they retained the vertical position proper to human
beings, whereas the philosophy of most of "ours" amounted
to no more than the formula: When in Rome, do as the Romans do.[7]

I would like to emphasize that what is being opposed here is not
nations (and, *a fortiori*, not races). The Western Belorussians do not
constitute a separate nation. They seemed to be foreigners because
then (in the '50s) they had one advantage over those from eastern
Russia: they had not yet been boiled in that pot in which peoples still
having some crystalline moral structure rise upward, while what is
left—the progressive mass—leaks out at the bottom. This process
of the loss of prejudices (and, at the same time, of conscience) has
been expressed in the saying "Where the conscience used to be,
a horseradish has grown."[8] And, as we know, a horseradish is not
a great moralist. . . . No natural force can implant a soul in a lump
of flesh controlled only by conditioned reflexes. Even if Martians
were to come and take away the present administration, the
horseradish that has grown where conscience used to be would not
become a moral principle. . . .

But perhaps one can mentally distinguish between the people's
flesh and its immortal soul? Otherwise there simply are no people.
I have already said and written that there are no people. It was
objected to me that there is likewise no intelligentsia. The arguments were persuasive; but for all those arguments, I directly felt
the reality of the intelligentsia. What kind of reality? I don't know.
I just feel how it flutters. Apparently the same thing holds for the
people. In the popular mass something quivers, and crawls out to
take the form of petitions for the opening of churches, of sectarian
communes. . . .

[7]Boris Khazanov, *The Smell of the Stars* (Vremya i My Publishing House, 1977), pp.
280–81.
[8]There is a suggestion here that "horseradish" is a euphemism for "prick."

The horseradish has a firm grip on its own truth: not to piss against the wind. But the soul is a dubious reality not confirmed by science. It just exists somehow and somewhere. And whoever begins to believe in it and behave absurdly, bunching off like the sectarians and the dissidents, is looking for trouble. So that to go with the soul of the people means to go against the popular mass. And conversely, to go with the popular mass means to trample the soul of the people underfoot.

If the inner voice is very strong, a person feels it sooner or later. Perhaps not right away. Perhaps only by the age of thirty or forty, on the basis of bitter experience. But he will certainly feel that to reject the role of renegade means, under certain conditions, to accept the role of scoundrel. He who has understood this will not forget, although in what words he will express his experience I do not know: words may vary. And if he looks around at others, it is only at people with a conscience. He is not frightened by the life of a renegade—the break with the mass. Confucius said that when virtue rules, it is shameful to be far from court. But when vice rules, it is shameful to be near the court. I think that here the word "court" can be replaced by the word "people." The sense is not changed. Heaven can turn its back from the people just as from the state and the court, and then to be a renegade is not at all shameful. It is merely difficult.

It is especially difficult to live as a half-renegade, half cut off from the mass, not breaking one's connections with it, not closed up in one's own sect. When the mass is in one of its bestial states, to go away from it; and when it becomes more human and human questions are stirring in it, to reply to those questions without hatred or rancor. The mass does not change, but some people hear and come closer.

To live as a half-renegade is to sit on two chairs. This is sometimes possible (when the chairs are together), and sometimes impossible (when the chairs separate, and one's behind falls into an empty space). In such a case one must firmly decide where to sit. There is a time to live, and a time to die. There is a time for compromises, and a time to refuse compromises. Either you are ready to become a renegade, you permit yourself that position when the mass becomes bestial, or you come, like Shchedrin's liberal, from agreements "as possible" to agreements "on something, at any rate," and you end up "in conformity with baseness."

The present-day liberal looks around at others as he lives. He has no commandments, and the categorical imperative is given as an idealistic eccentricity. But there is shame and conscience. That is enough, until he has broken off from friends and been dragged into compulsory association with that which has grown where conscience used to be. With the anticonscience. The liberal huddles himself up and bristles, but he can't go off into himself, close himself up within himself. He has no depth in which he can keep silent. He wants to talk, he wants to find understanding. And they offer him understanding: Just recognize the truth of the horseradish. Realize that anticonscience has its own sincere convictions, its own reasons. And a person understands. He gets used to a dialogue with the anticonscience. And how such a dialogue ends, we all know. And then it is hard to go back to what you were before.

We are living in a society where one cannot insure oneself against violence. And the example ascribed to St. Augustine is applicable to all of us. I have already cited it in "Letters on Moral Choice," but I shall cite it again. It is pertinent here.

When the barbarians captured Hippo, many virgins were raped. Augustine considered guilty those who experienced—to use the terms of this article—agreement with the anticonscience, who experienced momentary satisfaction from their complaisance. Those who felt nothing but horror, pain, and revulsion had not sinned.

I think that Augustine's model can be applied to the victims of any violence. Savonarola recanted under torture; but when the torture was stopped, he repeated what he believed. The violence mastered his flesh, but it did not master his soul. Such a soul remained pure.

The power of violence may last even longer—not for minutes or hours but weeks and months. In my view, that too is forgivable, if the fainting of the soul stops together with the circumstance of complete powerlessness, solitude, and desperation. One must not judge severely a person fallen into conditions beyond his strength, even if other people could have borne up under it. One must not judge one person by a yardstick suitable only for another. Let him judge himself: we won't cast a stone at him.

But now a month, two months, of freedom have passed. The fainting of the soul is over. If the soul has remained alive, it comes to itself. And if it doesn't come to itself? If the person, even in freedom, continues to mutter what he said out of fright?

I have cited the examples of Galileo and Uriel Acosta. But Galileo did not write a statement protesting the anti-Catholic campaign of the foreign press. And Uriel Acosta did not become an assistant of Ben Akiba.

What can we call a person who actually likes his complaisance in a dialogue with the anticonscience? And who, of his own free will, continues what was begun under lock and key? That person's soul was ready for a new role. Here violence played the role of a midwife, and helped a true understanding of its nature to be born. And the sympathy we felt for the victim disappears. We are not about to call complaisance toward the anticonscience a new kind of martyrdom.

Perhaps my words sound too sharp. I do not insist on them, and am prepared to close more gently—with the words of a poet:

> And for you, flying in bad times,
> Under the whip of war, after the power of a few,
> At least the honor of the mammals,
> At least the conscience of the pinnipeds!
>
> And the sadder, the more bitter for us
> That men-birds are worse than animals,
> And that against our will we believe more
> In vultures and in falcons.[9]

[9]From Osip Mandelstam's poem "Again the Dissonance of War," 1923.

VICTOR NEKRASOV

A Strange Man

I belong to that rare category of people who do not like—who are even a bit afraid of—famous persons. Either I am timid with them, or I'm afraid of seeming more stupid than I really am, or I blurt out something that will make me blush for an age afterward. In short, I avoid them. And now I can't forgive myself for never having got to know—and I could have, I really could have—either Boris Pasternak or Anna Akhmatova. (I met her for the first and last time at the St. Nicholas Cathedral in Leningrad, when she had gone to her eternal rest.) Nor did I approach Mikhail Zoshchenko, although I was present on that memorable evening at the Writers Union on Voinov Street when he, agitated and stammering, read his stories to some Leningrad writers who were no less agitated than he was. That was a year and a half before his death.

In short, I am not drawn to great men. I'm afraid of them.

But when, quite unexpectedly (although of course after a preliminary telephone call from Moscow), there appeared at the dinner table in our home in Kiev an academician who was shy, taciturn, and (most important) didn't drink (the other guests weren't used to that, I must confess), I could hardly believe my eyes. Moreover, I was somewhat put out because my wife had to heat up the two little pieces of herring that she had procured with such difficulty and had prepared with such effort.

66

"Andrei Dmitriyevich doesn't like anything cold," his wife, Lucy, said, spreading her hands helplessly. "There are no scientists without quirks. Even his kissel has to be warmed up. And the balcony has to be closed off."

I closed off the balcony. What else could I do?

Yes, Andrei Dmitriyevich has many quirks. Not only herring, kissel, or complete perplexity at a railroad ticket window, where his card as a Hero of Socialist Labor (three times!) could solve all transport problems in a moment. No doubt he has two or three dozen other quirks. But there is one that the people who consider themselves the leaders of our country can't get used to—simply can't understand. The man fears nothing. Nothing, and no one.

Bravery, valor, fearlessness, gallantry, heroism? No, none of those fine, lofty concepts applies to Sakharov. I believe that in him that feeling—the feeling of fear—has simply atrophied. Maybe he just doesn't think about it? He lacks the time, even for other, more important things. People, people, people. Fate . . .

I would like to count myself among Sakharov's closest friends, but I don't have the right. We have seen each other only a few times, and we are of different casts of mind. (My usual "Without a stiff drink you can't make sense of anything" is alien to him, alas!) Moreover, I am not distinguished for any special ambition or vanity, and yet . . . I am infinitely proud (I emphasize those two words) that the noblest, the purest, the most fearless, the best, and probably the most learned (here I'm not qualified to speak: in my school years I had to have a tutor in physics) man has a favorable attitude toward me, and even forgives certain sins.

I am also proud of the fact that I—I alone in the world—have a photograph of Andrei Dmitriyevich that I personally took in a hospital in Moscow, a photograph that has not appeared in any *Life* or *Paris-Match* or *Stern*, and never will. I alone have it. It stands on my bookcase. In Sakharov's way, a bit embarrassed, it smiles at me. When I wake up in the morning, it's the first thing I see. And I somehow feel warmer . . . because I not only love that great, strange man, I don't fear him.

V. TROSTNIKOV

The Death of Ivan Ivanovich

> I had a home country
> Of beauty supreme.
> There the spruce swayed above me;
> But it was a dream.
>
> —A. N. PLESHCHEYEV
> (1825–1893)

Ivan Ivanovich was not simply an executive, *he was an executive who had felt out the astute line of behavior*. And what did that consist of? I shall explain right away. Let's say the head of a section in his department comes into his office with a document he has asked for. Ivan Ivanovich does not look at the document, as most executives do, but at the face of the man who has come in. From the expression on his face he can tell whether the man is expecting a dressing-down or praise. That comes easy to him. But it is handled as if only a deep familiarity with the substance of the case made it possible to make the right decision. As a result, even when the interview turns out to be a dressing-down, the visitor leaves the office with a feeling of respect, and sometimes admiration.

Having mastered these fine points of his work, Ivan Ivanovich would have risen, ultimately, to a very high post—the kind he didn't even dream about when he was young. The next level in the hierarchy was that at which you were mentioned in reports of receptions or meetings at the airport. True, they didn't mention you by name but only in the phrase "and others." Still, it was plain to see in the photograph who was among those "others." Yes, that promotion was quite possible and even very probable. They talked about it at the top, as

68

Ivan Ivanovich's friends confidentially told him. But everything turned out differently.

He somehow didn't think about it, while the years went by one after another. And lately they had begun to fly. Then, imperceptibly and little by little, he reached a kind of boundary line and felt that he was not what he should be. The first bell had rung.

In what did his state consist? To put it briefly, in the appearance of *indifference*. He began to feel indifferent toward everything: toward money, toward food, toward resorts, toward tourist trips abroad, and even toward the good intentions of his bosses. There were, I should say, only two things that still gave him real pleasure: sports broadcasts on television (especially soccer) and walks along the river with his dog.

His physical health was still good, even enviable. His heart rarely troubled him, and he had no pains in the liver, kidneys, or stomach. He still had strength in his muscles, and a stamina that was rare for his age. He slept soundly and peacefully. But within him something had changed, had become dead and cold. Life had become uninteresting, even burdensome. One evening Ivan Ivanovich caught himself entertaining the preposterous thought that it would be good to go to sleep right away and never wake up again.

It all began with a chance meeting when, as it happens, he was walking his dog. An old fisherman, unhurriedly applying tar to his flat-bottomed boat, told Ivan Ivanovich how he had twice been a prisoner of war in Germany, in 1918 and in 1944. His first imprisonment was something he remembered with satisfaction. He lived as a laborer on the farm of a rich peasant. He had more than his fill of sausages, and drank milk instead of water. His second imprisonment was horrible, and he escaped only by a miracle. When he and his fellow prisoners were being transported from one concentration camp to another, he asked them to shove him through a small window under the roof of the cattle car. He was terribly thin at the time. They pushed, and he fell out on the other side. It was night, and the train was going full speed. Naturally, he broke a few bones, but he survived. He made his way along, moving stealthily at night and hiding in the woods or underbrush in the daytime. When he reached the Ukraine, he hid himself among his own people.

And so on—the story is understandable. Ivan Ivanovich had heard
of such things before. But this time he did not take it all in in the
usual way, and for a long time he could not forget the story. What
amazed him was not that people had experienced such things, but
something else. The old man's calm, good-hearted account, without
the slightest complaint or feeling of offense, with no call for vengeance, compelled him to take a new look at the life of the people
and see there a stratum he had not noticed before. From that time
on, he often thought of the old man. And he thought not mentally,
not logically, but with his heart and instincts, as it were. And that
stratum began to strike him as increasingly important and a serious
matter.

That disturbed him. All his life he had put Party guidelines in
first place, and suddenly it turned out that there was something still
more important. He asked himself: "What is it?" But he could find
no intelligible answer. He just felt that it was a kind of genuine,
rooted life of the people in which harshness and goodness were not
different, opposed principles but two sides of one more fundamental
thing which softens harshness and makes goodness inevitable. It was
the kind of life that isn't a bed of roses. A serious life, always burdensome, but which cannot be lightened once and for all, since then it
would cease to be serious.

That elusive but very important reality unexpectedly appeared in
certain memories associated with his uncle from Vitebsk, with his
Aunt Marfa, who had come from Mikhnev, with a near-fantastic peasant hut, with a Russian stove heaped with sheepskin coats and felt
boots. And although he could not understand where he had seen those
things and when, they somehow became most important, as though
if they didn't exist, nothing would exist at all.

Somewhere close to the level of clear consciousness he now experienced an inexplicable sadness which made itself felt at the most
awkward moments, and then came embarrassments.

One of those embarrassments occurred on a hot July day when
Ivan Ivanovich, accompanied by the secretary of the regional
committee, the oblast architect, and two industrial consultants,
visited a state farm where the building of a slaughterhouse was
planned. They looked over the site, checked it against the blue-

prints, discussed details of the plan, and then their work was finished. Their car still wasn't back: they had dismissed the driver for three hours, and had managed to complete their work in two. In the administration building it was stifling, buzzing with flies, so they went outside and sat on a bench in the front garden. Suddenly Ivan Ivanovich thought longingly about cold milk from the cellar, and decided to go in search of a hut with a cow. Near the ruined church with its tent-shaped roof he saw a well. An old granny was hauling up water in a bucket. He went to help her, and carried the water to her hut, which was not far away. He followed her into the yard, put the bucket down, and asked for a jug of water to drink.

"Go inside and rest, and I'll pour you some right away."

"Don't you have any milk anywhere?"

"What milk, dear? There's only one cow for the whole village."

He went into the spacious hut with timbered walls, and his gaze was immediately drawn to a huge ikon, about a yard high, in a thick wooden frame and under glass. If he had seen anything similar, it had been only in the Tretyakov Gallery or in the Novgorod Kremlin. The Mother of God, depicted at full height, in profile, was walking rapidly, so that her clothes fluttered. From the Byzantine face and the dark colors it was plain that this ikon has been painted long ago.

"Where did you get that image?"

"Why, from that same church where we got the water, the Beloved of God. That's what the church is called, the Beloved of God. That's church ikon. It was hanging at the entrance. In '36 they closed the church, and I hid it. But now, you see, I'm not afraid. I hung it in the room. Maybe they won't touch it now."

"But it's worth a lot. I think a museum would be glad to buy it. Haven't you asked anybody?"

"Of course they'd buy it! But I won't sell it to a museum. Priests and other people from the Petrovsky Church have been here to see me. 'Give it to our church,' they said, 'so people can pray to it.' They even offered money."

"So why didn't you give it to them?"

"Ah, my son, it mustn't leave the village. If it did, the village would get sick and fall to pieces. And maybe they'll open our church sometime. Then how would we manage without the church ikon? It should stay with us."

Ivan Ivanovich looked at the ikon again. The Mother of God was walking across our land and blessing it . . . yes . . .

When he had thanked the old lady for her hospitality, he left the hut. The shortest way back to the administration building was past the backyards. Beyond the gates the land was overgrown with drop-wort and leonurus. Unmown hay! But there was no one to mow it, and there was no reason to. There were no cows. The old granny had said the village would get sick. Lord, wasn't it already sick? Where had the past gone? The sturdy farms, the haymaking, the night watch of horses put out to pasture, the village round dances? Again he remembered his grandfather and his Aunt Marfa, and his heart was heavy. After all, he was one of those who had fiercely destroyed all that. Atrocities, concrete-block construction in the villages.

But it's not that which depresses me, he thought. I shouldn't pretend that I feel sorry for the peasants. I feel sorry for myself. Never in my life will there be a quiet stream with water lilies and duck-weed, the sound of women washing clothes by beating them with sticks on a planked footway extending into the stream, the fusty smell of hay. All that has passed me by, never to return. And what has there been in my life? The eternal cigarette smoke of conferences, eternal intrigues and scheming, unofficial telephone calls from the top and to the top. And then the pseudo-democratic jargon in which the Party members speak to one another, and which is pure hypocrisy, since there are all kinds of intonations and nuances that immediately indicate to the initiated the precise place of each one in the hierarchy. They are all supposedly on familiar terms with one another, and all are hail-fellows-well-met, but everyone knows his place. In general, instead of beauty and freedom, fate has given me pettiness and servitude. But the whole thing is, that I myself chose my own fate.

He went up to his colleagues, who had been waiting for him, and that is when the awkwardness hit him. A ruddy-faced consultant with thick, moist lips and reddish curly hair, obviously trying to please him, said cheerfully: "I think you were that hut's last visitor. As soon as construction of the slaughterhouse is begun, all this is going to be cleared away."

He made a sweeping movement with his arm, taking in the area where everything would be cleared away. The area included the old church; and in its very center, so it seemed to Ivan Ivanovich, was the Beloved of God, blessing our land. Without saying anything, he rapidly turned away and began to cough very hard. Then he went over to the fence, as if looking out for the car.

Of course his colleagues at the office couldn't understand the reason for the strangeness in his behavior. If he had explained to them with complete frankness that he had suddenly been overwhelmed with the unsatisfactoriness of the life he had lived, they would only have laughed and said: "All right, Van Vanych, you were only joking. Now tell us the truth. Give us the real reason."

After all, any of them would have sold his own father to achieve what he had achieved. Nonetheless, they all felt that something was wrong with him, and all talk of promotion ceased of itself.

It all ended the way it had to. One day he was summoned to the Center in connection with some contrived business, and when it was settled they went to see the chief. The Party chief—the same man who had backed his last appointment—lit up a cigarette, gave one to Ivan Ivanovich, slapped him on the knee, and said: "You're tired, Van Vanych, very tired. You've given a lot to the job. And you've done so much, that a man simply envies you. Just remember . . .

And then he began to talk about their joint work in the region, then in the Center. He recalled instances when Ivan Ivanovich had shown unusual cleverness in getting out of difficult situations. His elevated tone, one much like that used in speeches at funeral feasts, affected both of them, and they became tearful. But in his heart of hearts Ivan Ivanovich knew very well that everything he had done was pure fiction; that all the difficulties he had learned how to get out of so skillfully had arisen only because of the gap between Party instructions and real life; that they had all been spinning around in an artificial space they themselves had created; that their efforts, far from being useful to the basic life of the people, had constantly damaged it. He knew that if that stratum of real life had been preserved, it was not thanks to their activity but in spite of it.

They put him on the retired list with a maximum of benefits: a top pension with the right to use a government dacha for the rest of his life. And that dacha gradually gave him tranquillity and cured him of his apathy. That first summer, everything that wanted to grew on his plot of land: a full-branched balsam with little yellow flowers, garden angelica as high as a man, and golden clover. But the next year he cleared the land for useful crops. For two seasons he took an interest in strawberries and made lots of preserves that he even served to his neighbors. But then his calculations led him to the thought that it would be more profitable to raise flowers, and he plunged into that new activity. He had to read special literature, and go to experienced people for consultations, but it paid off. In the spring he had tulips, in early summer peonies, then carnations, and in autumn dahlias and chrysanthemums. All those flowers had to be sold. And at that point he developed a real taste for making money. At first he sold everything at wholesale prices to a woman he knew who carried the flowers to the city markets. But in time, overcoming his embarrassment, he took a stall for himself, because the woman was keeping too much of the profit, and he couldn't allow that. He grew more and more stingy, and finally began to think about using every square centimeter of his plot of land. Some of the trees growing were casting shade that was harmful to the flowers, so he regularly poured acid around them; and when they died, he got permission to cut them down.

What his new phase of inner evolution would have led to is unknown, because one fine day he had a heart attack. While waiting for the ambulance, his relatives laid him down on a sofa, fussed about, and oh'ed and ah'ed. And the last thing he heard anyone say was: "We have to use Uncle's identity card to buy things at the special store before they find out he is dead."

But those words didn't offend him; they brought no pain to his heart. He didn't even understand their meaning. He was already far away from where his body convulsed with agony.

He went out from the cool shade of the edge of the woods onto a field brightly lighted by the slanting rays of the morning sun glittering with thousands of little puddles not yet melted after the night's cold. He walked toward the other end of the field, but not directly: he kept changing his direction, so as to step with his full weight in the very center of the nearest glazed-over puddle. And when he did that, a sound like that of a Hawaiian guitar would begin at the edge and radiate to the center. It would grow rapidly in volume, then end in a pleasant crunch. And at that moment, his foot would sink down several centimeters.

As for what awaited him at the other end of the field, none of us may know. The fate of his soul is a great secret which no one can look into here on earth. But Inna Lisnyanskaya's poem tells the fate of his ashes:

Over there, beyond the fence, their fists unclenched,
Row after row, row after row, the old Bolsheviks sleep.
Above them, neither aspen, nor birch, nor alder—
Only posthumous sorrows and immortal sins.
And government tombstones like serried ranks. . . .
Lord, have those made in Your image come to such misfortune?

RAISA ORLOVA AND LEV KOPELEV

The Sources of a Miracle

Andrei Sakharov's appearance was a miracle.

He was the youngest member of the USSR Academy of Sciences. Deeply involved in complex problems of physics, he was respected by his colleagues and by the authorities. He was the recipient of government prizes and, on three occasions, of the nation's highest award, the Gold Star of a Hero of Socialist Labor. His future looked as serene as his past. . . .

And then suddenly (suddenly to an outside observer) he turned off the beaten path and began to defend persecuted and unjustly convicted people. Crimean Tatars not allowed to return to the Crimea. Germans not allowed to go to Germany. Jews not allowed to go to Israel. Russian Orthodox and Catholics, Baptists and Pentacostals, persecuted for their beliefs. Workers oppressed by their superiors. And he strove for political amnesty and the abrogation of capital punishment.

He came to the Writers Union when Lidia Chukovskaya was expelled. When he received a telephone call telling him that someone's home was being illegally searched and he couldn't find a taxi, he hitched a ride on a truck crane going in that direction. In Omsk, Mustafa Dzhemilev was being tried. The police forcefully removed both Sakharov and his wife, Elena Bonner, from the courtroom corridor. In Vilnius during Sergei Kovalev's trial, once again Sakharov was standing at the door. Likewise in Kaluga, when Alexander Ginz-burg was being tried. And in Moscow, when Anatoly Shcharansky was on trial. This kind of thing happened many times. . . . And although he had already had a heart attack, he went to Yakutia to visit an exiled friend. There he and his wife walked twenty kilometers through the taiga.

He was called in for questioning by procurators and officials of the Academy. They warned him. They tried to dissuade him. They threatened him. His apartment was broken into by drunken hooligans, Palestinian terrorists, and certain "relatives" of those who had died in a subway explosion, shouting that he was defending the murderers. He received telephone calls and anonymous letters threatening to kill his children and grandchildren. Manuscripts were stolen from his desk. And finally, without a trial, he was exiled to Gorky and put under house arrest, with a whole subunit of uniformed policemen and plainclothesmen standing guard.

But he does not give in. He continues, again and again, to defend human rights, to call for justice and political common sense.

Many people forget in their admiration for Sakharov's feat about the very deep tragedy of his life, and about the deadly threat to which he is subjected. Sakharov's fate is tragic because his soul is torn between a passion for science ("More than anything in the world I love radiation . . . ") and a love for people—not for mankind in the abstract but for suffering, downtrodden individuals.

He is very ill. He lives under constant nervous tension. With each passing day the danger to his physical existence increases.

His enemies cannot understand him. They call him a foolish eccentric. But there are many more who see him as a holy ascetic.

One often hears people ask: "In our country at a time like this, where did this incomprehensible, inexplicable miracle come from?"

Well before the world knew about Sakharov, he opposed ministers, marshals, and Khrushchev himself in insisting on stopping nuclear tests. He spoke out against the charlatan Lysenko when the latter was all-powerful.

He gave all the money from his government prizes, more than 100,000 rubles, for the construction of cancer hospitals.

Why is he like that?

Andrei Sakharov is unique and, as a genuine miracle, cannot be fully explained. The genius of a scientist is a gift from God. One can, however, try to trace the sources of his world view, the sources of that moral strength which has made him a spiritual leader, the personification of the best hopes of contemporary Russia.

From early childhood, Andrei Sakharov breathed the air of the Russian intelligentsia. Beginning late in the eighteenth century, there were several generations of rural priests in the Sakharov family. Andrei's great-grandfather, Nikolai Sakharov, was an archpriest in Arzamas honored by his parishioners for his goodness and enlightenment. His grandfather, Ivan Nikolayevich Sakharov, was the first not to follow the career of a priest. He became a lawyer and moved to Moscow. Early in this century he was editor of a collection of articles titled *Against Capital Punishment*. He was an acquaintance and collaborator of the writer V. G. Korolenko. A friend of the Tolstoy family, the musician A. Goldenweisser, was Andrei Dmitriyevich's godfather. His father, Dmitri Ivanovich Sakharov, became a physicist. He was also a talented pianist.

With the first fairy tales told him by his grandmother, with the sounds of the piano played by his father, with poems and books, Andrei took in that spiritual culture from which grew his notions of good and evil, of beauty and justice.

We have often heard him recite Pushkin by heart, almost to himself: "When for mortal man the noisy day is hushed . . ." He once said: "One would like to follow Pushkin's example . . . genius can't be imitated. But one can follow something else, perhaps higher. . . . "

We were talking about how delighted Pasternak had been by Camus's Nobel speech, and Andrei Dmitriyevich remarked: "That's like Pushkin. That's Pushkin's code of honor."

He and his brother Yuri would recite Schiller's "The Glove" with youthful excitement, interrupting each other. And they remembered a game they played in childhood: One would "mumble" a rhythm, and the other had to guess what poem of Pushkin's the first was thinking of.

We met Elena Bonner and Andrei Sakharov one evening in 1971 at a poetry reading given by David Samoilov at the Writers House. Since then we have often read the poems of Pushkin, Tyutchev, Aleksei Tolstoy, Akhmatova, Arseny Tarkovsky, and Samoilov together, and listened to the songs of Okudzhava and Galich.

But it was not just the spiritual traditions of the past, not just literature, that formed Sakharov's world view. He was a son of his time. As a schoolchild, as a student, as a young scientist working on the atomic bomb, he believed in the ideals of socialism; he believed in the just greatness of his country. But precisely because he believed deeply, sincerely, and purely, he perceived all the more sharply the gulf between the ideal and reality and, as he matured, experienced all the more painfully the collapse of his youthful beliefs.

In 1978, in an interview with *Le Monde* around the tenth anniversary of the Prague Spring, he said that the decisive turning point in his destiny had come at about that time. It was in July 1968 that he first published his memorandum on the peaceful coexistence of the two social systems.

One hundred years ago, in his speech on Pushkin, Dostoyevsky said: "To be a real Russian means to be universally human." Today that is being confirmed again by Andrei Sakharov.

SEMYON LIPKIN

Chapters from the Novella
Ten Years

A Chronicle

Chapter Four

Amirkhanov had explained it correctly to the general. The mountain
village of Kurush had actually got its name from an ancient Persian
king who in the Russian textbooks is mistakenly called Kir, which
provokes amused perplexity from Persian scholars listening to Russian
historians at symposia, since the Persian *kir* corresponds in both sound
and sense to our shortest swearword. But who had given the village
its name? The territory of the present-day Gushano-Tavlarsky Autonomous Soviet Socialist Republic had been in a state of vassalage under
King Kurush. Was it true that the mighty conqueror himself had visited that wild spot in the mountains and, amazed by the village tow-
ering like a solitary monarch over the crests, had named it in his own
honor? Or had his descendants done it? Or his attendants? The present writer is merely someone fond of reading historical books, a dilet-
tante, and is not in a position to answer that question. When, rather
recently, in the eighteenth century, there suddenly appeared on the
land of the Tavlars the upstart Shah Nadir—a man who, like Hitler,
was insolent and not very literate—he tried to prove that the chunk
of land he had seized belonged to Persia, citing the fact that the highest mountain village in the area had a Persian name. But that was the
same kind of nonsense that was recently spread by the Germans sur-

rounding Leningrad. This land, they said, is German, and the proof lies in the names of the cities: Petersburg, Peterhof, Oranienbaum. What do we know about past centuries? What do we know about past years? The textbooks lie, and the newspapers lie. Only the myth is true.

The village might have been named Kurush by the Gushans, who are linguistically related to the Persians, and who had complex relations with the Achaemenid dynasty; but it is known for a fact that the Gushans never settled in those areas—they didn't go that high up into the mountains. And the height can drive one crazy. Just past the district center, rising almost vertically above the hilly kolkhoz pasturelands, is a narrow path about a yard wide (sometimes a yard and a half) which extends upward for a kilometer. Once when he was young, the present writer climbed up that path to the village, and his heart sank with fright at the unbelievably deep abysses on both sides of the stony path. Twenty years later he once again crossed the hilly pastureland to the path, but he couldn't bring himself to climb up to the village. He couldn't even believe that once he had dared to do it. And how ashamed he felt when he saw schoolchildren, laughing and leaping, running along that fearsome path, which moreover was slippery, because it was then late autumn.

That almost vertical narrow path between the abysses connected the inhabitants of Kurush with the rest of the world, which they called "the lower world." Their land was barren, divided into strips so small that, as they said, you could cover them with a felt cloak. The men would take off for six months at a time to earn some money. Some would travel to nearby places, reaching the Don; others would go farther. They would go to Turkey, Syria, and even Egypt, where, it was said, one man from Kurush became a vizier. The men from Kurush engaged in various crafts: the craft of blacksmith, the craft of goldsmith, the craft of beggar. And some of them were very much feared by the children in the nearby villages. The children feared those who engaged in the craft of Moslem circumcision, detecting them from far off by some kind of instinct.

Fate had taken the blacksmith Ismail farther north than the others: he had helped build the Volga-Moscow Canal. Once in 1932 he spoke rudely to a tax collector who had been harassing him, and the

latter, even though he too was a Tavlar, brought criminal charges
against him for unlawfully slaughtering his own sheep, and Ismail got
five years. (The very fact that it was such a short sentence showed it
was a trifling matter.)

By the time the great disaster overtook his people, Ismail was
past sixty. He had seen a lot, and he knew a lot. He read Russian
and Arabic; he had visited the Cossack camps on the Don and in the
Kuban; and he had worked in the forges of Damascus, where the
best steel in the world was first made. But he had never seen anything like what went on during the building of the canal. "Doomsday!
Dadzhzha"—the Moslem Antichrist—"has come!" exclaimed his fellow villagers when, after returning home, Ismail told them about the
corpses floating between the dams.

Ismail had come back lame: his leg had been crushed by a rock.
His sentence had been reduced to three years, not because he was
lame (even though lame, he had been kept on hard labor) but because
he had earned extra credits. Often he had overfulfilled the norm 150
percent.

Oh, how they celebrated in Kurush! In a Russian village, people
would probably have been afraid to give such a warmhearted, even
rapturous, welcome to a former zek. But all the inhabitants of Kurush
were of the same clan, the same blood, and community is higher than
the state, more important than the state, more solid than the state.
Even the cliffs, the clouds above the cliffs, and the bushes among the
cliffs were of the same tribe as the people; and they too took part in
the village celebration.

And how it warmed Ismail's heart when, after the experience of
the canal—heavy as fetters riveted to your feet—he again saw his
wife Aisha, older and thinner, his son Murad, his helper in the blacksmith's craft—now a well-built adolescent with a hooknose like a bird
of prey—and his friends! He saw, too, his native rocks, clouds, trees,
the huts with their lean-tos: his native village, surrounded on all sides
by a vertiginous abyss, with only the path, thin as a ribbon—the
Lord's bridge to paradise connected with the rest of the Tavlar earth.

Ismail again became the kolkhoz blacksmith. During the first years
after his term in the labor camp, he was helped by Murad. Then
Murad was drafted into the army. Ismail's only remaining close rela-

tive was his sister Fatima, who had married a man in a village farther down the mountain. Amirkhanov, the district secretary, of course knew about her older brother's past, but he had no choice but to take him on, since most of the men were away at war. Also, Fatima was of the poorest class, and she and Ismail, along with her husband, had been among the first to join the kolkhoz. Ismail taught his sister how to read and write Russian, which at that time was a rare thing among the mountaineers. And Fatima was a hardworking, clever woman, a leader among the farm workers. True, she had something to hide: she was religious. But on the other hand it was precisely because of that that the other kolkhoz members respected her and believed her.

When, on the eve of Lenin's Day, Ismail, smelling of smoke, with his neatly trimmed beard covered with soot and his eyelids red from the fire of the forge, limped back to his hut, his heart was filled with joy. His favorite nephew, Fatima's son, Alim, had come up from the lower depths to visit him so that they could be together on the holiday. The thirteen-year-old boy had already propped up his paintings in crudely knocked-together frames against the wall, beyond the hearth, on a plank trestle bed. Old Aisha, using what was left of the barley, had managed to cook some unleavened flat cakes for her nephew in the ashes of the hearth. Also, she had set some nuts aside for the winter (a nut tree grew in front of the hut), and now they gleamed golden on the small round three-legged table.

Ismail and Alim embraced, but, as is the custom among Moslems, their lips didn't touch. The boy had not yet learned the eastern way of not letting excitement show on his face (it was a rather long face with big eyes); and he was excited because his uncle began to study his paintings, like a master looking over the work of a master. The paintings were copies—portraits of the leaders, and portraits of more attractive persons. Karl Marx looked like a Tavlar mullah, except that he lacked a turban. Ismail liked the portrait of Aisha, which was a full-length one. The nephew had prettied up the blacksmith's wife. He had depicted her in a rich, flowing shawl (which she did not have); and on her legs and feet he had put short red stockings and red shoes (which she likewise did not have). Ismail also looked approvingly at the portrait of himself. Alim had done a pencil sketch of his face and part of his torso, cutting it off at the shoulders of his beshmet. Ismail

was amazed at the resemblance, not understanding that the young
artist had failed to catch the expression of his penetrating blue eyes.

"Why paint a bad man?" Ismail asked reproachfully, pointing with
one soot-blackened finger at a portrait of Stalin. The boy's mouth fell
open in holy terror.

Aisha, disapprovingly shaking her head, on which she wore a black
fillet, said: "The Prophet forbids painting."

"The Prophet forbids painting Allah," the blacksmith objected
confidently, "because no one and nothing is hidden from Allah, but
he himself is hidden from everybody and everything. But not a single
sura or verse of the Koran says that the lame blacksmith Ismail and
his old woman Aisha can't be painted."

In Kurush the appearance of a new person who had come up from
the lower depths, even if that person was a child, was always an event.
One after another, the blacksmith's neighbors came to his hut. They
expressed their delight at Alim's pictures, snapping their fingers and
using their tongues and lips to make the kind of sound used in Russia
to drive horses. Then they reluctantly went away. One of the visitors
was the one-legged, one-armed Babrakov, wearing the medal and ribbon of a wounded soldier on his field shirt. He was an important
personage, the head of the club. With his good arm he embraced
Alim as an adult. Then, leaning on his crutch and on the boy, he
sighed in the Moslem way; that is, giving the sigh a definite meaning.
When he had sat down on the trestle bed, he said in a didactic tone:
"Never forget, Alim, that on your mother's side you are from Kurush.
This is your homeland. So give our club the portraits of the leaders.
And your mother will be glad for her son, for her kin, when all of
Kurush, that minaret of the mountain country, looks at your pic-
tures."

Having said that, Babrakov again sighed significantly. Ismail understood that the head of the club wanted to communicate something
important to him, and he waited for him to begin. Babrakov began:
"The tongues of our women are like millstones. The mill grinds away,
but there is no grain. Didn't you hear anything, Ismail?"

"What can you hear at the forge? The bellows swell up, the fire
leaps, the iron rings."

"You're wise, Ismail. But today the roar of the forge isn't stopping you from hearing, and tomorrow it won't stop you—it's a holiday. Prick up your ears; we need your advice. As for you, Alim, I've been thinking it over. We won't take your pictures as a gift, we'll buy them. We'll make it official. Maybe, if we can, we'll pay you in food products instead of money."

"Food products are better," Ismail answered for his nephew. "You said something about this and that kind of talk. We all know that thistle grows on the cliff, while rumors grow in the public square. When we go to the club, we'll listen and we'll find out."

The village club stood in the middle of a broad and unevenly graded public square on a gentle slope of the mountain. It had formerly been a mosque, and nothing in its structure had been changed except for two square apertures cut into the walls for showing films. Those apertures destroyed the intricate ornamentation of the walls. No one in Kurush—not even Ismail, who read Arabic—knew that the ornamentation actually consisted of words using the ancient Arabic alphabet, and that the words were taken from sayings in the Koran. But under the Russian slogan "The cause of Lenin and Stalin will triumph," learned Arabists could have read the eternal words of God and His Prophet saying that one must fear fire made by the infidels—fire whose fuel was people and stones. And yet, although the mountaineers knew neither the ancient nor the modern Arabic alphabet, they had spared the club while ruthlessly destroying nearby buildings, because the club had formerly been a holy mosque.

The villagers had already assembled in the square. They knew that after the speech, a film called *Lenin in October* would be shown. And although they had all seen it several times before, their lives were so dominated by hunger, poverty, and boredom that it was pleasant to wait for some kind of entertainment. The older women wore their hair bound in fillets. Over these they wore threadbare flowing black shawls folded into triangles, with the ends hanging down their backs. The young women and girls were dressed in a more modern, city style, but some of them still wore the high, cylindrical hats decorated with embroidery and a silvered ball. Despite the winter, the war invalids were in field shirts without felt coats, and had worn mountaineer's shoes on their feet. But all of them wore huge sheepskin caps on their heads, because a mountaineer may have a ragged beshmet but he must by all means wear a good sheepskin cap (the best thing about a man is his head) and carry a knife. (Unfortunately, knives were prohibited.) The boys also wore sheepskin caps and tattered, ill-fitting jackets with felt hoods. Like gray eagles on mountain cliffs, the eighty-year-old men sat tailor-fashion.

Ismail shook hands with all the men. Then he was approached— and this was a breach of custom—by Sariyat Babrakova, the kolkhoz shepherdess. She was a tall woman of about thirty. Her eyebrows met above the bridge of her nose like a stripe of black paint; and her face, with high cheekbones, was tanned by the wind. She smelled of snow and sheep urine. Her first husband had been killed at the front, leaving her with two children. She had taken as her second husband the one-legged, one-armed head of the club, Babrakov, who had come back from the war more than six months before, but the marriage had only recently been registered. At first her neighbors had disapproved, but by now they had calmed down. It seemed that only a few months had passed, but it was already evident that Sariyat was expecting a third child.

Once when she had been herding her flock in a pasture higher up the mountain where the grass was thicker, she had been attacked by a wolf. The wolfhound couldn't cope with her attacker, and Sariyat killed the wolf with her heavy staff. But before she did, he had torn the felt coat that had belonged to her first husband, and she had had to patch it, somehow, with pieces of felt. (Before the war, women had never been shepherds.)

In her husky voice, which was not like a woman's, she said: "I want to ask you, Ismail, to give me a straight answer. After all, you're one mountaineer who can read and write. And you spent three years near Moscow itself, though not of your own free will. You know the lowlands and the highlands. So tell us, Ismail, why is it that today a different man is going to speak to us about Lenin?"

"What different man?"

"Who has made the speeches to us during all the years of the war? Fazilev, the editor of the district paper, has been sent up to us. But today a different man is eating and drinking at the chairman's home. He's soon going to show up here. And do you know who that new man is? Biyev."

"Biyev? The chief of the district NKVD? Mr. Need-Better?"

"The chief of the NKVD is going to tell us about Ilich any minute now. Some boys saw him going into the chairman's house. He has a big belly—bigger than mine—and he has a revolver on each side."

Ismail remembered the anxious, incomprehensible words of Sariyat's husband, Babrakov. Yes, it was a time to keep a clear head. It wasn't the business of the chief of the district NKVD to make a speech about Ilich. Ideology was not his field. The chief of the NKVD had other business.

As though reading his thoughts, Sariyat added huskily: "There's a rumor that they want to drive us out of Kurush to a lower village. They'll repair the ruined houses and settle us in them. And it's hard for the big boss to come up to us. So they sent Biyev so he can get us ready in advance for the resettlement, and at the same time he's going to be in charge of the Lenin evening."

"Allah Akbar, Lord of worlds, what will happen to Kurush? What will happen to the graves of our ancestors? Can the living leave their dead behind them forever?"

Such were the questions Ismail asked. He asked himself, and he asked the villagers standing around him. A young man in a wheelchair came rolling up, smiling maliciously. He was handsome, a true Cherkess, as though he had leaped from the pages of a Caucasian poem by Pushkin or Lermontov. But now he couldn't leap from any pages, since he had paid with both legs for his Stalingrad medal.

"*Salaam aleikum,* Ismail."

"*Vaaleikum salaam,* Akhmed. Any complaints about your wheelchair?"

Ismail had made Akhmed's wheelchair, which he had even supplied with handlebars for greater convenience. The blacksmith had thought up the design himself.

"My tank is in good working order, thank you. Well, today we're going to get instructions from Biyev as to how we should proceed down the mountain. Get your kitbag ready, Ismail. And our Kurush . . ."

Akhmed didn't finish. Biyev had appeared, accompanied by Babrakov. Everyone noticed that the chairman of the kolkhoz wasn't with them. Why not? Didn't he respect Lenin? On either side of

Biyev's soft belly was a black holster. Under his arms he carried the portraits of Lenin and Stalin drawn by the boy Alim.

The villagers crowded into the club and sat down in the former mosque. So that everyone could see them, Biyev placed the portraits of the leaders right on the stage, in front of the table, and then sat down next to Babrakov, who was seated behind the table, having propped his crutch against the back of his chair. Another portrait of Stalin hung in a niche which at one time had indicated to praying Moslems the direction of Mecca. Kuchiyeva, the Party secretary who had the same last name as Ismail, climbed up the long-unwashed wooden steps to the stage. Babrakov, after some brief introductory remarks corresponding to the occasion—a sad one, but a cause for great optimism—announced that our esteemed Comrade Biyev would make the speech.

The chief of the Kagar District NKVD was a tall, heavy-faced, big-bellied man. His head was set on his shoulders as though he had no need of a neck. His little pink eyes were bloated. The villagers of Kurush had preserved a very pure Turkish speech (true, there were many words of Arabic origin, but they had taken on a Turkish sound with the accent on the last syllable); and Biyev, although he was reading from a typescript, spoke Tavlar badly. Also, if he departed from the typescript, he combined Tavlar words with distorted Russian ones. He was used to concluding his speeches with a shouted appeal: "Need better!" Once he had declaimed: "Long live the soldiers of Dzerzhinsky, our organs of security, who are doing a good job of serving the Soviet Union! Need better!" Since then he had been nicknamed "Mr. Need-Better."

He said little about Lenin, talking more about Stalin and the imminent victory, which required efforts and sacrifices. The speech, as usual, tied in the great causes of the whole country with the concerns of the kolkhoz and the tasks of Kurush. Having shouted with genuine enthusiasm all the necessary toasts, already incomprehensible, and having waited until all the necessary applause died down, Biyev, disregarding his typescript once and for all, announced: "The portraits of the very great leaders of all peoples were painted by a fifth-grade student, Alim Safarov, who is here among us. Well painted! Need better!"

There was another burst of applause, this time wholehearted and approving. Alim plainly felt shy. Everyone noticed this, and the applause increased. But Biyev held up his hand to indicate that he had yet another announcement to make. The villagers sat down again, and listened.

"For a long time the kolkhoz members have been complaining about how hard life is in Kurush. And they have had good reason to complain. You don't have a doctor. No medical worker, so to speak, wants to come up here. You have no school. No teacher wants to live under such conditions. In general and on the whole, the Party oblast committee and the government of the republic have considered the complaints of the kolkhoz members and have decided, even though it is wartime, to improve your life, to give you well-built houses in one of the lower villages. Now things will be good for the children. There is a school there. In Kurush there are quite a few war invalids and sick old men and women who need medical care. Soon Sariyat will present us with a little bundle—a future dashing horseman—and she won't have to climb down the path with her pack load. There is a hospital nearby. Dear mountaineers, I congratulate you! Get ready for a new life. Need better!"

With his joke about the pregnant Sariyat the NKVD district chief had intended to evoke a gay liveliness among the audience, and to show his understanding of ordinary human anxieties and joys—something that always brings one closer to the people. What he evoked instead was fear, dismay, indignation, and abusive language. And something unforeseeable happened: Alim went up on the stage, seized the portraits of Lenin and Stalin, and, holding them high, leaped down rather than coming down the steps.

"We'll never leave Kurush!" shouted the villagers. "It's better than all the towns of the lower world! We'll never leave the town of the dead—the graves of our ancestors!"

Akhmed, in his wheelchair, rolled up to the stage. Leaning on the handrail, raising himself up on his stumps, legless and handsome, he shouted into Biyev's fat face words that were both strong and powerless: "What you're saying is bad, Biyev! It's lowdown! Are you a mountaineer? You're a fat swine. May the infidels eat you!"

The woman Sariyat, pregnant with a "specially resettled person,"

rose to her feet. She rose, big as a cloud in her shepherd's felt coat.
"Cursed be the womb in which you were conceived, you pig-faced
devil! Where is the chairman? Why is he hiding from us?"

Russian obscenities, mixed with elaborate Tavlar swearing and
Moslem curses, shook the walls of the former mosque. Not staying
for the movie, everyone went out into the public square. Biyev and
the Party secretary slunk away unnoticed. No one knew that both the
Party secretary and the chairman of the kolkhoz, warned by Biyev,
were now busy packing up their things. They had been authorized to
take not one but three loads for each member of the family.

Biyev was anxious and nervous as to whether, at his house, they
would manage to pack up things in time. He had the enviable right
to take food and other things without limit. But his wife had no sense.
The only hope lay in his mother and mother-in-law, practical old
women. Unfortunately, he himself had to stay in Kurush until the
next morning.

The people of Kurush didn't sleep that night. For a long time
there was shouting in the public square. It was suggested that those
who knew Russian well, Ismail and others, together with the wisest
old men, should write a letter to the oblast committee and the Council of People's Commissars in the name of Devyatkin and Akbashev,
who, although he was not from Kurush, was a Tavlar, and from the
Kagar Gorge at that. In his high post, his Tavlar heart could not harden.

The stars shone, big and low; and the mountain peaks, lulled by
the music of their shining, went to sleep. But the villagers in their
huts did not sleep. How could they leave the place where they had
lived from time immemorial? Where they had lived back when there
were no Moscow bosses? When there was no Moscow? How could
they leave the minaret of the mountain land? Alim had read somewhere that Kurush was the highest populated place in Europe. And
when would the resettlement begin? Probably not before summer,
because they first had to repair the ruined houses below. Ismail mentally composed a letter, but he knew the idea was useless. As one
who had helped build the Volga-Moscow Canal, he knew the big bosses
very well.

The villagers went to sleep just before dawn, but at dawn they
were awakened. Douglases droned over the mountain peaks, and

dropped paratroopers on the uneven terrain of Kurush. Young Chekists broke into homes, demanding that within an hour the people pack up their things, one load per person, including children. Biyev and the commander of the paratroop detachment divided the detachment into groups, with two paratroopers in each. They had calculated things so that there would be twice as many paratroopers as homes. Semisotov knew how to count.

There were also women among the paratroopers, and not just because men were more needed at the front. The humane government had realized that the operation was an unusual one. Among those being resettled the majority were women: quite a few tottering old women, quite a few that were sick, and possibly some pregnant ones. And for them a "frail" woman Chekist was more suitable than a heavy athlete.

Two paratroopers, a young man and a girl, broke into Ismail's hut. Both were snub-nosed, smooth-faced, and seemingly without eyes, since no soul shone in their eyes. What showed there was something dull, not even bestial, but a kind of malice alien to everything living.

At first, the two of them shouted and swore. Then they cooled down and even began to help gather things up so as to speed everything up. But they did it in such a hurry! Finally, three loads were packed up. Alim had put a saddlebag on his shoulders. Under one arm he held the portraits of Lenin and Stalin, and under the other those of Ismail and Aisha. Plainly, he had decided to leave Marx behind.

The girl paratrooper was exasperated. "You motherfucker, why did you take five things? You were told in plain Russian to take one load per person. You're a stupid kid to take the pictures. Maybe there are better things here. But they have to be left—that's the order."

"I painted them myself, and I'm not going to leave those portraits!" Alim shouted. "You can kill me, but I won't leave them!" In his shout one could hear both the tearful tone of a child and the anger of an adult.

The other paratrooper said in Ukrainian: "Let the kid take his paintings, Polina, and when we get to the truck we'll see. They won't let him put the paintings in the truck."

"Okay, take them, you motherfucker."

The villagers assembled, every last one of them, as Semisotov had ordered. The crying of children, the cursing of women, the terrible silence of the elders, and the even more terrible, tragic silence of the mules with their beautiful eyes. They started down the path. Every group of five persons was escorted by a paratrooper. Biyev led the way, and the commander of the detachment brought up the rear. On that almost vertical, precipitous path the Chekists lost their self-confidence. They felt dizzy on that little thread between two abysses.

Ismail was carrying the heaviest of the three saddlebags. He had of course realized—even before dawn he had realized—that it was not a question of resettling the inhabitants of Kurush in a village farther down the mountain. In that case, they would have waited until spring, or even summer. Biyev, the district security chief, had lied. All the people of Kurush—perhaps the whole tribe, the whole republic—were being resettled in far-distant places. Maybe in Siberia? That was why Biyev had deceived them: he was afraid of the resistance the villagers might put up. (But why be afraid? All of them had long since been bent like horseshoes.) That was why they had been ordered to take only one load per person. And that was why the paratroopers had been dropped on Kurush.

Nor was it only Ismail who realized the enormity of the catastrophe: others did, too. Wasn't that why, when they reached the halfway point on the path, all of them, as though by agreement, pausing to catch their breath, looked up and back for a moment? Their homes were no longer visible. Only the minaret of the agricultural club, like a solitary, daydreaming pilgrim on the way to Mecca, stood still in aloofness and reverence. The sky was aflame with the dawn, and two-headed Mount Elbavend loomed up clearly. One head seemed to crown the trunk of a body crucified by the morning sun. Heavy, icy eyelids drooped on the other, wrapped in a snowy turban. The mountain did not want to see—could not see—the great grief of its kinsmen. The end of a people? A people being driven away?

Long afterward, in a distant foreign land, that moment would continue to live in the hearts of the people. But now the moment passed, and they started down the mountain again. It seemed to Ismail that it was hard for his nephew, who was walking in front of him, to carry the saddlebag, along with two paintings under each arm. He wanted to lighten his nephew's burden, and he tried to take from him at least two of the paintings, but his lame leg gave way. Ismail fell. The paratrooper right behind him couldn't help him in time, and the old blacksmith Ismail Kuchiyev lost his footing on the path and fell into the abyss, being smashed to bits at the bottom. Lenin and Stalin, too, fell into the chasm, as did the legless Akhmed in his wheelchair made by Ismail. So also did the one-legged, one-armed Babrakov, with his bundle and his crutch, along with several old women and children. The commander of the detachment felt disgruntled: the number of resettled persons would not correspond to the number on his list. Moreover, one of the paratroopers lost his footing as well and fell into the chasm, all because of those traitorous mountaineers, the black-eyed hirelings of Hitler.

But the mountains stood there, watched, remembered, and wept— they wept with the never freezing tears of springs. No, those tears will never freeze. The paratroopers will die, and the children of the paratroopers, and the grandchildren of the paratroopers. But the mountains will stand there, thinking, remembering, and weeping; and on those wrinkled faces the springs of tears will never dry up.

Chapter Five

The train left Alma-Ata in late March. Moving along the Turksib, it reached Arys on schedule. There it was warm, and the *dzhida* had already begun to flower. Then the train began to go more slowly, plainly not in a hurry to get from Asia to the north, to Moscow. It spent almost the whole night in Kzyl-Orde, standing for hours at stations big and small, and even out in the grassless steppe, like a sick man with angina pectoris in the streets. On the seventh day it reached Ruzayevka. There, at that junction, it mindlessly maneuvered for a long time, finally stopping on a siding along the track leading farther north.

The lights of Ruzayevka shone through the cold twilight. To get to the first track, to the station building, one had to make one's way across the platforms of the cars of other trains, and sometimes virtually under the wheels, and go clear around the long, tightly sealed hospital train. As usual, many of the passengers, both military and civilian, carried teapots and mess tins. The train was overcrowded, a great many passengers having accumulated. During the boarding at Alma-Ata there was such a crush that the women conductors, to make things easier for themselves, had closed the doors on passengers, both those with tickets and those without them, and even on generals and colonels. But the men of lower rank proved to be cleverer. Many of them had provided themselves with special handles made by skilled craftsmen among their fellow soldiers with which they could easily open the rear doors of the railway cars.

In Ruzayevka the soldiers and officers rushed to the commandant's office to get what food they could, using their food chits; and only one officer went to look for mail. They sent off a telegram: because of the train's long delay it was being held up. His orders instructed him to reach his unit on the same day that the train finally got to Ruzayevka, and the officer of course did not know when his trip would end, all the more so since he was planning on spending two or three days at home in Moscow. From the mail office he went to the commandant's office. When the train had been approaching the station, it had seemed to him that many lights were shining; but it turned out that the station was enveloped in darkness, snow, and mud. Everywhere people were bunched together in almost hopeless waiting, talking in Russian, Ukrainian, even Polish. The officer stood in line. When, after forty minutes, he reached the window, he told the deputy commandant that he hadn't brought a food chit with him, but that he was hungry and wanted a special chit for a loaf of bread.

"I can't do it without a regular certificate, Comrade Captain," said the deputy commandant in a bored voice.

But the captain, having acquired experience from others, knew how to answer. "I'm sorry, Comrade Senior Lieutenant. I was given only five days for a visit with my family. I didn't have time to get the form filled out, because I wanted to get to the front as quickly as possible, and the train has been moving very slowly. I'm very hungry."

He could not get food on the basis of a regular chit because in Alma-Ata he had got everything a month in advance and given it to his father. The deputy commandant, sitting there in his rear-area safety, ill-humoredly issued the captain a special chit for bread and a package of food concentrates. The captain found out, however, that to get them he would have to go rather far: to the very end of the station and then, going out into the city, across a square.

At the station, bartering had already begun. Soldiers were trading the nuts and raisins they had got in Alma-Ata, along with bits of clothing (mittens, for instance), for moonshine vodka. They were trading with Mordovian women, arguing with them, and demanding to taste the vodka. Where the pitted, dirty asphalt ended and the last streetlamp no longer shone, there stood a cattle train. Three soldiers and a sergeant in sheepskin coats and felt boots stuck into galoshes were showing the way to officers and soldiers with special food chits: after the third car, turn to the left, and you will come out on the public square.

Suddenly one half of the side of the second cattle car was pulled back, and in the opening the captain saw a young woman in a white smock. The sergeant helped her jump down to the ground, and asked: "What's going on, Zinka?"

"Wait till I catch my breath. It's a premature birth. The foolish woman couldn't wait until the right time. But those people are healthy as horses. Even though she couldn't wait until the eighth month, the baby is okay. He won't die, so he'll live."

"Who are those people?" asked the captain, not hoping to get an answer, since he realized what kind of forces these soldiers belonged to.

But the sergeant apparently thought there was no point in being secretive: "They're not people, Comrade Captain, they're traitors— families of Vlasovites. Shameless renegades, you might say. From the Caucasus, I guess."

"Can I take a look?"

"Why not? Go ahead and look. But not for long. You'll find it disgusting. They're savage, they stink from the rear end, and they've got fleas."

The captain peered in. The railroad car, intended to haul cattle, had been reequipped to haul people, but in such a way that it was worse for the people than for the cattle. Along both sides of a narrow passage were plank beds. Neither above nor below could people sit up straight. They were bunched together in that garbage heap, in the filth and stench. The former shepherds had become flocks and herds.

A toothless old man in a sheepskin cap, sitting on the floor of the cattle car, which was filthy, covered with spittle and hardened excrement, eagerly breathed the damp air coming in through the open door. In a corner the newborn baby was crying. Women surrounded the mother. Men with many days' growth of beard sat on the plank beds—silent, immobile, and fierce. Their bare feet were waxen, like those of corpses.

"To think that once these people were pink-cheeked babies in the arms of their mothers," the captain thought, irrelevantly recalling Annensky. The features of these unfortunate people seemed strangely familiar to him. He leaned in and said: "*Salaam aleikum. Khardan siz? Kim siz? Tavlar?*"

"*Tavlar, tavlar,*" confirmed the men, showing white gums in a smile.

The captain didn't know enough Tavlar to continue the conversation in that language, so he switched to Russian: "Why are you here? In this cattle car?"

In reply came shouts in women's, children's, and old people's voices. "We're cattle. We're food for the Russians! They're resettling us! They're sending us to Siberia! They're resettling our people! Who are you? Are you from our part of the country?"

"Are you in your right mind? Are they resettling a whole people?"

"They're resettling all our people! That Georgian dog Stalin is resettling us!"

"Is Musaib Kagarsky among you? And even Akbashev? And all the others? And the Gushans?"

"They left the Gushans there. They left them and our dead ones. Musaib is here, and Akbashev is here, but they're riding in good railroad cars. While we, as you can see, are worse than cattle. Time was, when an ewe in the flock would have a lamb, we would take care of both the mother and the child. But now the woman Sariyat has given birth, and there is the breath of Allah in her and her baby boy, but there is no water for her."

"Do you have a bucket?"

"Yes, but they won't let us out to get water."

"Give it to me, and I'll bring you water."

The captain thought that the Chekist sergeant would be angry with him, but he turned away. Perhaps he turned away on purpose. In a Russian, anger flares up, but it cannot burn goodness. Goodness is not wood, not coal, not kerosene, but the spirit of God.

The captain had earlier noticed a spigot with hot water. He hurried to it, mixed hot water with cold, and returned to the cattle car. A boy whose bloodless face seemed to be all eyes took it from him without thanks.

The captain went to get some food with his special chits. He was issued a loaf of bread with an extra piece, and a concentrate of kasha. He ate the extra piece, and the bread turned out to be sour.

When he came back to the cattle car, the door was closed. He asked the sergeant to open it for a moment, saying he just wanted to give the people inside some bread and kasha, but the sergeant refused. "It's not allowed," he said, adding quietly: "It's an order, I got bawled out."

In some confusion, the captain headed for his own train, not sure he was going in the right direction. He made his way across the platforms of passenger and freight cars, and went around silent locomotives.

The captain was named Stanislav Yurevich Bodorsky. He was a poet and translator, and since the beginning of the war he had worked on the army newspaper *Son of the Fatherland*. When the front moved to the west and the army was for some reason left in reserve near Proskurov for reforming, Bodorsky received from Alma-Ata, where his parents had been evacuated, a telegram saying his mother had died. The newspaper's editor, Colonel Emmanuil Abramovich Prilutsky, refused his request to go on leave to attend the funeral: a soldier, he said, must overcome personal grief. But a member of the military council who had taken a liking to the captain and recently given him the Order of the Red Star sympathized with his army writer and authorized him to spend a furlough of five days in Alma-Ata, giving him ten days in all for the trip.

Now Bodorsky was returning. He was late, because the train had barely crept along, but he hoped that the army was still near Proskurov. If not, he would find it. To get up to an advanced position is always easy.

He had been amazed by the resettlement of the Tavlars. As usual in a trying situation, he thought first of all of himself. Even when he learned that his mother had died, he thought first of himself. But no one should be judged hastily. Turgenev, who described the execution of Tropman in detail, turned his head away at the last moment. Having read Turgenev's article, Dostoyevsky spitefully noted: "A frightful concern, to the last degree of scrupulosity, about oneself, about one's own safety and one's own tranquillity—and that in view of a lopped-off head!" Nonetheless, the present writer considers Turgenev not only a great writer but a good man.

Bodorsky's name was associated—at least in the eyes of the literary establishment—with that of Musaib Kagarsky. Bodorsky's father, a gray-mustached Pole, short, with a shaggy head set low on his athlete's shoulders, and with bright-blue eyes under beetling brows, had in the past been a gendarme officer. Probably that was what prompted Stanislav's two older brothers to join the Communist Party. Both had fought in the civil war. One of them had been killed in battle near Sinelnikovo, and the other had vanished in 1937.

Stanislav himself took after his mother, an Armenian from Yelisavetgrad. He was tall, slim, swarthy, and black-browed. Unlike his brothers, he kept his distance from the regime, not even joining the Pioneers.

His poems were far removed from all the prevailing trends in Soviet poetry. Stanislav's idols were the Symbolists—especially Sologub and Vyacheslav Ivanov. As he saw it, they possessed everything he was striving for: spiritual tension, elegance, heavenly music, mystery. The Soviet poets—both the Proletarians and the Formalists, both those on the left and those on the right—repelled him with their pragmatism, their dependence on current circumstances, their verbal poverty, and their search for a *point d'appui* outside of poetry.

In 1926, when he finished middle school in his native southern city, he went to Moscow, almost without money, in pursuit of two goals: he wanted to try to publish his poems in the capital; and he wanted to get a job as a worker at a plant so that after he had earned some seniority (while of course concealing his father's past as a gendarme), he could get into the university. Editors refused to accept his poems, saying they were melancholy in an old-fashioned way, and that he should frequent literary circles and learn from Demyan Bedny,

Zharov, Bezymensky, Utkin, and Molchanov. But he did succeed, despite the widespread unemployment, in getting a job at the Derbenevsky Chemical Plant. His work, harmful to one's health, involved distilling metanilic acid for azo dyes.

Stanislav rented a corner of a two-room apartment in a wooden building in the Malaya Tatarskaya district. The apartment had low ceilings, and the toilet and water hydrant were outside. The landlady worked with him at the plant. Her husband was a watchman who worked for twenty-four hours at a stretch, and then was off for two days. When the husband wasn't there, Stanislav slept with the landlady.

She was an ugly, full-breasted woman with thick braids. She hated her husband, and would say to Stanislav in a singsong voice (she was from Poshekhone): "He's a relative of mine. His first wife was my first cousin once removed. I registered for residence with them, and came to work at the plant. Then my cousin up and died—she had cancer of the stomach. After the funeral banquet was celebrated, he lured me into bed with him. What else could I do? Where could I go? We registered our marriage—he didn't deceive me. But he makes me sick. I lie with him like a log. But with you I'm all on fire. I love you, my black-browed little Circassian."

Little Circassian? His father had boasted of his old, noble Polish family. He claimed that they were the younger branch of the princely Bodorsky family who had owned almost half of Circassia. Stanislav was always surrounded by books (he had two hobbies: music and history), and he had learned that in the valleys and foothills of Mount Elbavend there lived a tribe of Gushans; that under Gedimin one of their princes had become a Catholic, leading to the appearance of the princes Bodorsky in Lithuania and Poland. (The family name was derived from the name of the Gushan rulers' capital.)

Stanislav realized that he was in no way related to those Bodorskys—his father had lied. (The arrogance of the Polish nobles ran in his veins, but not their blood. He had been admitted to the petty nobility because he had attained the rank of officer in the gendarmerie.) But then, he could be pardoned his weakness. It was shared by some great men—Balzac, for example.

Meantime Stanislav, looking mockingly at himself in the mirror in the morning, would great his reflection in Polish: "Good morning, Your Grace and Radiance, Pan Stanislav!"

When he got his regular vacation from the plant, Stanislav went to Leningrad to see the Northern Palmira of Pushkin, Dostoyevsky, and Blok. He stayed with friends of his father. One fine day he found the courage to pay a visit to Sologub, not knowing, of course, that this was the last year of the revered poet's life. The door was opened by Sologub himself: bald, with a sickly, pinched face, a big wart on his cheek, and barefoot. Stanislav was too frightened to say a word. The two men stood there looking at each other until Sologub politely asked: "With whom do I have the pleasure of remaining silent?"

Sologub's apartment was big, empty, and cold. In the half-darkened study hung an ikon of the Mother of God. Stanislav recited about ten poems he had selected. During his recitation, Sologub kept nodding his bald head approvingly. But when he began to talk, he sharply—without raising his voice—reprimanded the young poet for his southern turns of phrase ("I like northern distortions, but I can't stand southern ones"), for imitativeness, for flabbiness, while at the same time noting some excellent lines. (That was just what he said: "excellent.")

Stanislav remembered that meeting for the rest of his life. At the time he was thirty-five, and had not yet published a single poem. But he was not, damn it all, a complete failure. He had managed, after working the required time at the plant, to matriculate in the history department of a teachers' college. When, in filling out the questionnaire, he had come to the question about his father's social status, he had put down the answer: "Civil servant." Nor was that untrue: at the time, his father was a low-level employee in the City Department of Municipal Services.

At the college, Stanislav struck up a friendship with a student named Daniyal Parvizov. It was the first time he had seen a Gushan in the flesh. They became especially close when Stanislav expressed the desire to learn the Gushan language from his friend. This was flattering to Parvizov, and he felt moved by it. This future secretary of an oblast committee was at the time the trade-union organizer for the class. He made arrangements for Stanislav to move from his dormitory on the Stromynka—a former poorhouse, where he lived in a room crowded with sixteen narrow beds and eight night tables—to his own spacious room, which was occupied by only four students, all of them Party members except for Stanislav.

Things went well. Stanislav acquainted the capable Gushan with the Russian language, Russian literature, and the culture of the Russian intelligentsia; and Parvizov was glad that his Russian friend was interested in the language, history, and folk poetry of the Gushans. Both were far from stupid. But although they lived in the same room for four years, hardly ever separating, both felt that the two of them were extremely unsophisticated and naive young men. Both were mistaken.

Stanislav once read to Parvizov a few of his poems. Parvizov didn't understand them very well because their language was strange—not the way Russians talk today. But he was moved by the very fact that Stanislav wrote poetry, since from childhood he had been accustomed to respect scholars and poets. From that time on, Parvizov became a kind of guardian for Stanislav, generously supplying him—in his capacity as trade-union organizer—with coupons for shoes, underwear, and once even an overcoat. He was aware that Stanislav did not look down at him as, for example, did the secretary of the class's Party cell, who—although with good intentions—always stressed the fact that Parvizov was a member of a national minority. But Stanislav made friends with him as an equal, with no attitude of superiority; and Parvizov, perhaps without being aware of it, was grateful to him for this.

He told his classmate about his people, about their ancient, mysterious past, and their fate. And once he sang in recitative a short folk legend and then translated it *viva voce*, using modern, colorless journalistic language. Stanislav was amazed by the great similarity of the Gushan tale to a Greek legend—of how Odysseus (among the Gushans the hero had had another name) cleverly deceived the Cyclops, blinded him, and made his way out of the cave enveloped in a sheepskin and mixing in with the sheep. Stanislav's naturally sharp ear caught the unusual rhythm of the tale—the voice from the remote centuries and the mountains—and he realized that he could reproduce that rhythm in Russian so that it would have a new, ringing sound. And so he translated the legend into Russian verse. Thanks to

a rough knowledge of the language of the original, he found turns of
phrase that, while correct in Russian, freshly re-created the sound of
the Gushan language. At the insistence of the delighted and exultant
Parvizov, he took the translation to the editors of a major journal, and
in a few months it was published. Not only that, but Gorky, in one
of his articles on the necessity of taking into account the multinational
character of Soviet literature, praised the translation of the Gushan
legend, even though he did so only in parentheses, and did not mention the name of the translator.

This was a success, an unprecedented success! Daniyal Parvizov
beamed. In a brief introductory note, "From the Translator," Stanislav had mentioned Daniyal as the author of the literal translation
he had versified. The names of the two friends both appeared in print
for the first time simultaneously. The student Stanislav Bodorsky became a Soviet poet, although of the lower—translator's—rank. And
when a new legend arose—about Musaib Kagarsky, illiterate but
wise—Gorky remembered Bodorsky, and on his recommendation the
unknown, budding poet was assigned an important state task: to
translate the orally created quatrains of the Homer of the twentieth
century, glorifying the motherland and Stalin, and castigating the
enemies of the people who had burned kolkhoz hay.

That August, when Stanislav and Daniyal, both of whom had
graduated from college, were spending some time in Moscow before
it came time to part, Stanislav received an invitation to Gugird, the
capital of the Gushano-Tavlarsky Autonomous Soviet Socialist Republic. So it was that the two friends set out together for Gugird—Parvizov forever, and Bodorsky on special assignment. At the Teplovska
station the car for Gugird was uncoupled from the express train going
to Baku and coupled to the end of a freight train, which, after making
its assigned run, was due eventually to come to a stop on a siding—
the Gugird station.

At Teplovska their reserved-seat car (there was no other direct
one) had been boarded by a young Gushan, who began to look for
someone. Seeing Daniyal, he began to talk to him in their native
language, but Daniyal gestured toward Bodorsky. The young Gushan
shook both of the Moscow poet's hands, and invited him into another
car. He began to help the confused Stanislav pack up his things. But

Stanislav asked if Parvizov, too, couldn't travel in the other car. The Gushan agreed, and took from the protesting Stanislav both his suitcases—one of them very heavy, with books.

The car that had been proposed to them stood at the end of the work train, the reserved-seat car not yet having been coupled to it. Daniyal was stupefied: this was the parlor car of Suleiman Nazhmuddinov, first secretary of the Gushano-Tavlarsky Oblast Committee of the Party. And Stanislav, too, was stupefied when the three of them went into the car. It had a kitchen and a living room/dining room, where an elderly Russian woman, very hospitable, was setting a table with dry wine, Dvin cognac, vodka, mineral water, and appetizers—sturgeon, caviar, cold chicken. Stanislav glanced behind the heavy portiere and saw a bedroom with two couches covered in brocade.

One of the bottles on the table contained a strange gray liquid. There was no label on the bottle, and Daniyal explained that this was *buza*. This was the first time that Stanislav had seen that beverage permitted to Moslems and mentioned by Pushkin, Lermontov, Tolstoy, and Bestuzhev-Marlinsky.

The young Gushan went into the bathroom. Stanislav immediately wanted to try the *buza,* and Daniyal was pleased by the Russian's desire to sample the national beverage. They drank a glass apiece, and Stanislav began to feel slightly tipsy. He suggested that they have another drink, but his friend stopped him. "We shouldn't. Let's wait for the representative of the oblast committee. His name is Zhamatov. The three of us will drink cognac to begin with. The reason you're being so heartily welcomed is that the word has got around: you were recommended by Gorky himself, the non-Party member of the Politburo."

In Gugird the friends took leave of each other. Bodorsky was taken to a hotel in a car. (It was his first ride in an automobile.) Zhamatov, the oblast committee's instructor for culture, apologized to Stanislav for the modest nature of the hotel accommodations, but the latter objected. And he objected quite sincerely: never since childhood had he been in such quarters. The suite consisted of two rooms, a bedroom and study. On the desk was a heavy, granite-like desk set, and next to it a telephone. On the table, sitting on a clay platter, was a huge round melon festooned with bunches of grapes. Next to it stood three bottles: again, cognac, vodka, and mineral water. In a corner—as in his childhood in their southern home—was a washstand of yellow wood and white marble. The bathroom and the shower, Zhamatov explained, were at the end of the hall. The windows looked out upon a luxuriant long park once planted by Prince Izmail-Bey, the prototype—so it was said—for Lermontov's hero.

Zhamatov asked permission to use the telephone. When he began to talk, Stanislav understood him: he was reporting to someone on the arrival of their guest. When he had heard the answer, he hung up and said: "Stanislav Yurevich, Suleiman Nazhmuddinovich has invited you to his home. Get some rest, and I'll come for you in an hour."

Stanislav washed up (there wasn't enough water in the washstand), unpacked his books and other things, put on a new pair of trousers and his only good silk shirt, and went down from the third floor to the street, having decided that he would wait for Zhamatov at the hotel entrance. Mount Elbavend was nowhere in view; and later Stanislav learned that the mountain's twin peaks could only be seen early in the morning, if there was no mist.

To the left, one-storied houses and clay-walled cottages stretched sadly away into the distance. To the right was a wasteland. Despite the heat, it was easy to breathe, and a lightly scented breeze came from the park. With their backs to the hotel entrance sat a stone Lenin and a stone Stalin—a marble version of a well-known and dubious photograph. From time to time, describing a half-circle around the statue, cars would drive up to the hotel.

Zhamatov, too, drove up, got out, and with a broad smile invited Stanislav into his car. They drove for a distance which, as it turned out, could have been covered on foot in five minutes. Stanislav could not understand why a car was needed. But he later understood why, just as he understood many other things in the behavior of the leaders of this little republic with its little capital: it was necessary to give a visitor the impression that the city was big.

The Party oblast committee had its offices in a three-story building built in the nineteenth century which before Soviet times had belonged to a wealthy breeder of astrakhan sheep. In the hallway, opposite the door, a soldier stood next to a table with a telephone on it. Zhamatov said to him: "We're here to see Comrade Nazhmuddinov."

The soldier nodded as if to say he already knew, and they slowly climbed to the third floor, not saying anything. Lord! He, Stanislav Bodorsky—only yesterday a student and poetaster with no reputation, and without hopes for one, alien to everything new, as if bogged down in the backwoods of the Silver Age—was now riding around in parlor cars and automobiles, occupying two-room suites at hotels, and would soon be received by a candidate for membership in the Central Committee, the first secretary of the Party oblast committee!

Zhamatov knocked on the white door of an office, and let his guest enter first. Suleiman Nazhmuddinov, legendary hero of the civil war, rose to greet them. He was very tall. His light-blue riding breeches puffed out above a pair of high boots, and his gray-green field jacket was of excellent fabric. His bald head seemed to be not so much bald as shaved in the Moslem fashion, and in need of a turban. His yellowish-red, sharp, predatory eyes looked over his guest as though he were prey. And even when Nazhmuddinov, according to the custom, asked Stanislav how things were going with his wife and children (which Stanislav did not have), he continued to resemble a gigantic bird of prey caressing a fledgling. Nazhmuddinov's gigantic body, his black handlebar mustache, the decorations on his field jacket (rare at that time), stunned Stanislav and gave him an alarming feeling of becoming acquainted with power.

The telephone rang. Nazhmuddinov put the receiver to his ear—too big and thick for his face—and began to listen. One could see that he was not pleased with the person he was talking to. "Comrade Professor, who told you that under alpine conditions you can't raise fine-fleeced sheep on a mass scale, averaging two lambs per ewe? Does science say so? While you were going to seven-year school, I became a shepherd. From the age of eight I carried a shepherd's staff and herded somebody else's flock. I herded them under the very clouds. You can't fool me! What kind of unique cases are you talking about? Listen, Professor! Tomorrow, *drën-matyr*, you'll be a former professor!"

Nazhmuddinov, obviously very pleased with the rebuke he had administered over the telephone, turned to Stanislav. He was like an actor satisfied with himself after having successfully played a short scene. He said: "We still have people who stick to the old production norms. We're against petty supervision, but what can you do with them? . . . I've heard you can speak Gushan."

Stanislav replied in Gushan: "I have a bad pronunciation. I'll never learn how to pronounce your t's and your three k's. And I don't have a very large vocabulary."

Zhamatov broke in enthusiastically: "What a pure pronunciation! A regular Gushan Djigit!"

Nazhmuddinov approved. "Constant attention to the culture of national minorities—that's what the father teaches us. You did a remarkable job of translating our folk legend. When I read it, I remembered my childhood, how my grandmother sang. We have many such legends. They're considered Greek, but they're ours. We're older than the Greeks, and we'll force bourgeois scholarship to recognize that. We'll force them. The father likes folk legends. I've read *David Sasunsky* and *Manas* of the Kirghiz. Frankly speaking, our legends are better. They're easier to understand. Do you know Tavlar?"

"No, I don't. I adapted Musaib from a literal translation."

"The Tavlar language is different. It's Turkish. Our Gushan language is more ancient. It's original. You should translate all the Gushan legends. It will make a big book. We have some prestige in Moscow, and a publishing house will make a contract with you."

"I'll be glad."

Stanislav really was glad. He had dreamed of such work. For him it was almost like his own poetry. He already saw himself as a second Gnedich.[1] No, more than Gnedich—a discoverer. He thought: I'll work all the harder at translating Musaib's stuff. It's published in *Pravda*. I have to gain the affection—even the love—of the leaders of the republic.

With a quick, professional glance he looked around at the study. The variegated furniture, the handsome bookcase, the bentwood chairs, and the comfortable armchairs that might have been found, say, in Chekhov's home. Only the desk was modern, ugly, and ponderous.

Nazhmuddinov had taken a liking to the young Russian poet who had been recommended by Gorky, and who knew Gushan, although

---

[1] A poet who translated the *Iliad* into Russian in the nineteenth century.—Trans.

not well. He said: "We'll make an agreement with the Writers Union—
with Shcherbakov. You'll be assigned to translate a narrative poem
that Musaib wrote on commission from us, *My Gushano-Tavlaria*.
We want to see it published in *Pravda* by the time the Stalinist Constitution is adopted. It's a poem about the happy life of the toilers in
the republic under the Stalinist sun. But in the first part there are
images of the distant and recent past, of our battle against foreign
aggressors. It describes how we voluntarily joined Russia."

"Voluntarily? But what about those long, bitter battles? In writing
about Shamil, Marx said that the peoples of Europe should learn from
him how to combat despotism."

Stanislav had not yet learned how to behave like a Soviet courtier.
But he would learn. Meanwhile, he found Nazhmuddinov interrupt-
ing him sharply: "Marxism, *drën-matyr*, is not a dogma. The Gushans
fought against the tsar but not against Russia. The Tavlars were back-
ward: they didn't fight against the tsar. Our older brother, the great
Russian people, saved our land from the grasping Persians and Turks.
Are you a Party member? No matter, you're a non-Party Bolshevik.
Some local historians, our people, helped Musaib, but he didn't understand everything. You'll have to bring the poem up to standard.
You'll be living at the old man's home as long as is necessary. We'll
make the arrangements. It's beautiful there in Kagar, and the air is
good. True, it's not very clean—it's not like in the Gushan villages—
but we'll take care of everything. Musaib is an interesting man, an
extraordinary man. I once paid him a visit. (I'm not an armchair type
of leader. Frankly speaking, I'm always with the people.) The old
man was warned of my visit. That was a mistake. He hid his clothing
in a trunk and put on a tattered beshmet, worn shoes, and a tattered
sheepskin cap. We show up, and we're met, *drën-matyr*, by a *di-
vana*—a foolish-acting beggar. I was ready to fly into a rage, but I
kept my self-control. (That's a land of ignorant people. As for me,
I love Dobrolyubov.) We sat down on a rug with holes in it, and
Musaib's old wife served us an omelet, sour clotted milk with garlic,
and nothing else. No *khinkal* and no wine. I began to shout at the
consultants who had come with me, and at the secretary of the Kagar District Committee. Imagine a poet known to the whole world,
whom Comrade Stalin knows personally, living in such poverty! I

gave orders to provide him with the proper clothes, to give him as much food as he wanted, and to report to me within two weeks. When I got back to Gugird, I was told that they had done everything I had ordered, but that Musaib had a trunk full of good clothes and footwear, and that he had also put his new acquisitions in the trunk. My consultants saw that on the underside of the trunk's lid he had pasted old pictures of the kind they had before the Great October Revolution: an advertisement for Vysotsky tea, a gypsy woman with a pack of cigarettes. Peasant psychology. But of course he's a great poet. We Gushans also have a poet who's not bad—a living classic, as good as Musaib, but very modest. His name is Khakim Azadayev. I'll introduce you to him."

Suleiman Nazhmuddinov indicated to Zhamatov that he should leave the room. Then the secretary of the oblast committee came around the desk, sat down on a bentwood chair opposite Stanislav, and fixed him with a piercing gaze from his yellowish-veined eyes, like those of a predatory bird. "Tell me. Who was Homer?"

The shock was such that at first Stanislav didn't understand the question. He almost decided that Nazhmuddinov was asking him about some Moscow Jew. But then he quickly realized that he meant the blind bard with whom Gorky had compared Musaib, and he answered accordingly.

Nazhmuddinov grew angry, but not at Stanislav. "Those damned consultants! They told me, dammit, that Homer was a classic of Marxism-Leninism. I objected to them: four!" And the former shepherd and partisan warrior spread his thumbs. "Four! Marx, Engels, Lenin, Stalin! Four!"

All of Nazhmuddinov's huge frame trembled with indignation, and he stamped on the long rug with his high boots. He was breathing heavily; and, as a curse, he spread out the fingers of his right hand. "Four! Four! Those damned consultants, the parasites, dammit! Four classics of Marxism-Leninism. Four, I told them. Where did they get a fifth one—Homer? Four!"

For a long time he couldn't calm down. Through the glass of the bookcase, Stanislav noticed an incomplete set of volumes of the Granat Encyclopedia which had apparently belonged to the building's former owner—the sheep-breeder. The volumes were scattered, and among

them was one marked with the letter H. Gesturing toward the bookcase, Stanislav said: "There's more there about Homer than what I
told you."

"I don't have time to read it, my friend. I just don't have time.
Yes, the Tavlar people have given us a great poet. And for your translation of the legend, a mountaineer's thanks. I read it, and I remembered how my grandmother sang that legend to me. I remembered
that one-eyed giant."

Stanislav did not have to be a genius to realize at once that the
Gushan Nazhmuddinov was not happy with the fame of the Tavlar,
but that he had to recognize it and bow down to it.

So Stanislav went to Kagar. And now, in the railroad yard, he told
himself that, God willing, right after the war he would tell, in chaste,
Pushkinian prose, how he lived for two months at the home of this
really talented self-educated poet. How he made a marvelous play-
thing out of the old man's poem, where lovely idioms and sayings
were concealed, as it were, by truisms. How *Pravda* published the
translation; how Stalin himself expressed his approval; and how he,
Stanislav Bodorsky, was admitted into the Writers Union. Then he
translated yet another narrative poem by the great Musaib, *A Song
About the Leader,* and everything else they made Musaib write: about
Pushkin's birthday, and about the fighters in the International Bri-
gades in Spain. But there was also joy: the publication, in Stanislav's
translation, of a book of ancient Gushan legends—a little bit of the
heart, one might say—with the brilliance of the versification and the
verbal archaeological excavations yielding the gold of ornamentation.
The book was a success not only with the state but among readers.
Many people wrote about it, including university scholars abroad.
Stanislav made money.

The narrative and lyric poems of Musaib in Stanislav's translations
were studied in schools. They were declaimed by children at holiday
soirées. People wrote dissertations about them. And at a Party congress, Sholokhov spoke to them as a major achievement of Soviet lyric
poetry in the civic vein.

Now Musaib, and all his fellow villagers and all his people, were being sent to Siberia in a cattle train. No. This mustn't be. He had to make one more attempt.

Carrying the brown loaf of bread and the package of concentrate, Stanislav began again to make his way across the platforms of the cars and between the wheels. He came to the first track, but the cattle train was no longer there. Had it been sent farther on its way, or switched to another track? Through the damp mist, the lights of Ruzayevka twinkled peacefully, since it was in the rear area. The railroad tracks scarcely glinted, then disappeared in the mist. Everything was silent: the station, the locomotives, the cars, the people. Like the damp, misty night, Stanislav's swelling grief was heavy and dark. He crawled onto the narrow platform of a freight car. Suddenly, the train started moving. Stanislav leaped aboard as it moved.

What Do You Think of Sakharov?

THE RESULTS OF A SELECTIVE SURVEY

GENRIKH: Okay. Now tell me the truth.

THE BURGOMASTER: Why, how can you say that, sonny? You're like
a little baby. The truth, the truth. . . . After all, I'm not just
any ordinary man, I'm the burgomaster. I haven't spoken the
truth to myself for so many years that I've forgotten what it's
like, that truth of yours. It makes me sick to see it—thrusts me
away. The truth, you know . . . what does the damned thing
smell like?

—E. SHVARTS,
The Dragon

There is no doubt that the "unanimous approval" of any political decision by "any Soviet government" is not so much an automatic reflex
proper to Soviet man as a function of the propaganda apparatus responsible for the matter in question. Any newspaper editor, any Central Committee instructor, can tell you without a moment's hesitation
what the Soviet people "approve" today, and what they "angrily condemn." And should any vagueness arise, he will pick up the telephone, call the right person, and then explain it to you. So we don't
need a Gallup Institute. To the contrary, any attempt to ask questions
of the public meets with hostility and suspicion. Why ask, when
everything is already known?[1]

[1] Soviet man is regularly called upon to express his opinion on only one "mathematical"
question: "What sequence of numbers will come up in the next round of the sports
lotto?"

And yet Soviet man, stupefied by propaganda and alienated from political decisions, remains a person. He thinks, he has his own opinions, and he frequently expresses them—in a waiting line, in a bar, in talks with friends or fellow workers. And although his opinions are a matter of indifference to "the powerful of this world," and the latter even try to pretend that they do not exist, it would be interesting to ascertain what the "talking production tool" does in fact think. We decided to try. By means of a selective survey, we tried to find out how the ordinary Soviet person feels about Andrei Dmitriyevich Sakharov. The answer to that question undoubtedly involves an evaluation of many processes taking place in the country.

Obviously, to obtain a representative sample it would have been necessary to question several thousand, or tens of thousands, of people chosen at random. As we know, however, questions about a "dangerous" subject are not popular in our country; and as a rule, one does not manage to get sufficiently sincere answers. Therefore, in lieu of a direct question we decided to substitute an informal chat containing leading, indirect questions. For example: "Have you heard that they've brought Sakharov back to Moscow?" Or: "They're expelling him to the U.S." Or: "They want to try him." The attitude toward Sakharov was evaluated by the interviewer, or by the interviewee himself, on the basis of a nine-point scale: from "an enemy, a spy, shooting him wouldn't be good enough" (one point) to "a hero, the conscience of the country" (nine points). In between: four points ("I don't know, but my attitude is unfavorable"); five points ("I don't know anything"); six points ("I don't know, but my attitude is favorable"); seven and eight points, two gradations of an increasingly favorable attitude ("useful," "very important"); two and three points, bad ("harmful to the country," "useless").

These conversations were conducted by fifty-three persons, each of whom gathered from one to forty answer-opinions. It is quite obvious that the sample obtained by such a survey is not fully random and hence is inadequately representative. The greatest danger arising with this method is a positive correlation between the opinion of the interviewer and that of the interviewee. Therefore, the interviewers were forbidden to question their relatives and friends, persons whose positive attitude toward Sakharov were sufficiently well known or could be assumed, and of course all dissidents or persons associated with them. This approach eliminated several thousand persons from the survey, but made it possible to get a quasi-random sample of persons whose opinion was not obvious in advance. True, there remained a certain danger that the interviewees (often fellow workers of the interviewer), whether deliberately or without thinking about it, would make their answer agree with the assumed opinion of their interlocutor. Therefore, the results obtained by those interviewers who had registered only positive answers were completely ruled out. Moreover, in the case of those interviewers whose results for a given social group of interviewees (blue-collar workers, students, etc.) proved substantially better than the average for the whole sample, the positive values were reduced by one point (eight instead of nine, seven instead of eight, etc.). That is, we assumed either that certain people, taking their cue from the interviewer, somewhat overstated their positive attitude, or that the interviewers shifted the scale upward by comparison with their colleagues. The correction did not change a positive evaluation into a negative one, but merely altered its value somewhat. All in all, about 16 percent of the material gathered was subjected to correction. The sample thus obtained can roughly be considered independent. As to the extent to which it is representative, I shall have a few words to say later on.

In addition to the interviewee's opinion of Andrei Sakharov, the interviewer recorded certain information about the former: whether a member of the Communist Party or a Party bureau, "nationality" (including mixed marriages), occupation, salary, age, education, place of residence, and sometimes certain special information, such as whether he was ever convicted by a court, did he have repressed relatives, religious affiliation, etc. These data made it possible to look at the results with a breakdown into relatively homogeneous groups.

The total number of persons questioned was 853: 470 men and 383 women. Their ages ranged from 18 to 78 years. Two-thirds were inhabitants of Moscow, and one-third lived near Moscow or in certain oblast cities in the European part of the country. Seventy-five percent of those questioned were Russians, 18 percent were Jews, and 7 percent were of other nationalities (Tatars, Armenians, Georgians,

Komis, Germans, Poles, people from the Baltic countries, and Ukrainians). One-third were engaged in physical labor, and two-thirds were white-collar workers. About 20 percent were members of the Communist Party.

These proportions differ markedly from the corresponding data for Moscow and for the nation, so that on the whole the sample was not adequately representative. Certain categories of the population were completely excluded from the survey: workers in police departments, and high officials of the Party-state apparatus, as well as high-ranking officers of the armed forces. At the same time, we did not take into account the opinions of persons known to have a favorable attitude toward Sakharov. This is a relatively small group, but it nonetheless consitutes not hundredths but tenths of a percent of the inhabitants of the capital. I judge from the fact that almost simultaneously with the survey—and in part by the same people—several thousand birthday photographs of Andrei Sakharov were distributed.

The fact that the sample as a whole was inadequately representative does not prevent one from trying to examine it part by part, and to answer the question as to what individual categories of the population of Moscow and other cities of the country (cf. the table) think about Andrei Dmitriyevich Sakharov.

This is, of course, not the opinion of the Soviet people; but one should think that the information furnished here reflects the mood of a not inconsiderable part of the inhabitants of our country.

Blue-Collar Workers

Interviews were conducted with a total of 245 blue-collar workers engaged in physical labor at industrial enterprises and at construction projects, at institutions, in the transport system, etc. (There were insufficient data for a breakdown into specialties and levels of skill.) The greater part of this group consisted of men with a middle-school or incomplete middle-school education, rarely with an elementary-school or middle technical-school education. More than half of them were young men under thirty. There were many members of the Komsomol, but the number of members of the Communist Party was relatively small (4 percent).

The attitude toward the questions asked turned out to be the most indifferent in this group. Two-thirds answered "I don't know," and more than half of the "I don't knows" were without any (positive or negative) emotional coloring. Those who expressed an opinion (the other third) were divided roughly in two: 16 percent were markedly unfavorable, and 13 percent favorable.

A positive or negative attitude was found most often among the men. Female workers constituted a noticeable share of those who answered "I don't know."

There was no marked correlation between the answers and age, education, and Party membership. There was, however, a reverse relationship (though not a very strong one) between the answer and the level of income. Among those who expressed a negative attitude, more were highly paid, while among those who expressed a positive attitude, more were low earners. However, the lowest-paid group— pensioners engaged in physical labor (as a rule, these were older, ill-educated people: female hospital attendants, charwomen, cloakroom attendants, nursemaids, etc.)—was distinctly unfavorable toward Sakharov (almost 80 percent) or neutral (18 percent saying "I don't know").

Engineers and Scientific Workers

Engineering-technical workers constituted almost 40 percent of the interviewees. These were men and women with a higher education, often Ph.D.s, both Party members and non-Party individuals. In view of the specifics of the survey, a large share of this group consisted of mathematicians and programmers. But it is hardly likely that the profession itself had, in this case, a noticeable influence on the answer. Sixty-one percent of the engineers expressed a definitely positive attitude, and among half of them that attitude was very positive. Only 20 percent were distinctly unfavorable. Among the men, the negative attitude was more clearly expressed, and corresponded well to Party membership. Among the women more vagueness was noticeable. Two-thirds of the answers were grouped around "I don't know," although "I don't know" with a positive attitude predominated. No correlation with age and income was observed in this group.

The Liberal Professions

This category included college professors, schoolteachers, doctors, journalists, artists, musicians, and others in the liberal professions. In this aggregate, the attitude toward Sakharov was even more favorable than among persons in the technical professions. About 50 percent expressed an attitude that was markedly positive, and another 15 percent one that was simply positive. Only 15 percent were distinctly unfavorable. Unlike the case with the engineers, the women in the liberal professions more frequently expressed a negative attitude than did the men. There was no noticeable correlation with age and income.

Administrators ("Leaders")

As has already been noted, Party and state leaders were not covered by the survey. We managed to poll only certain low-level Soviet administrators and scientists: directors of institutes, heads of laboratories, etc. These were middle-aged and elderly men, most of them Party members. Their attitudes were sharply polarized: two-thirds were decidedly against Sakharov, and one-third had a favorable attitude. Opinions among members of the Communist Party were divided in almost the same proportion: one-fourth expressed a positive attitude, and three-fourths a negative one—sharply negative in more than half of the cases. Almost all the Party activists—members of Party bureaus, secretaries, etc.—were in the latter group.

Youth

The attitude of youth in the negative part of the scale was rather close to the data for the blue-collar workers (15 percent "enemy"; 6 percent "harmful and useless"), while in the positive part it was close to the engineering-technical intelligentsia (34 percent "I don't know, but I'm favorable"; 24 percent positive answers). In this group, the percentage of those who replied "I don't know" was rather high: less than among the blue-collar workers, but more than among the other groups.

.   .   .

ATTITUDES TOWARD A. D. SAKHAROV
(Percentages of Those Questioned in the Group)

Designation (Composition) of the Group	Number of Persons in the Group	RATING BY POINTS								
		1 Enemy of the people	2 Harmful to the country	3 Useless, "Had it too good"	4 Don't know, but unfavorable	5 Don't know	6 Don't know, but favorable	7 Useful	8 Very important	9 Hero
Blue-collar workers	245	12	1	3	8	56	7	1	12	—
Engineers, scientific workers	331	6	9	2	9	10	33	11	19	1
Male engineers	172	8	16	2	6	4	28	20	14	2
Female engineers	159	4	1	3	12	16	38	2	24	—
Liberal professions	182	5	4	6	9	11	15	19	26	5
Administrators	46	40	6	22	—	—	—	12	20	—
Low-income blue-collar workers (pensioners)	49	—	4	14	60	18	—	2	2	—
Engineers, liberal professions (homogeneous)	415	9	7	5	11	15	25	12	13	3
Youth	295	15	1	5	1	20	34	10	14	—
CP Members	164	48	16	12	—	—	10	6	8	—

Thus, Andrei Dmitriyevich Sakharov provokes a sharply hostile attitude among about one-fifth of the adult population (20 percent of blue-collar workers, 21 percent of intelligentsia, 18 percent of working pensioners and other low-income categories).

The basic reason for the negative attitude is the traditional "He had it too good! What else did he need?" Often, an overall negative attitude was combined with an expression of disagreement with the measures taken against Sakharov. (A KGB executive said: "It invites comparison with Angela Davis, who, after killing a sheriff, came back to the U.S. as a professor. But he didn't do anything, and they exiled him. If he wants them to let him go abroad, then he should go.") But one often encounters much more bloodthirsty and less logical utterances. (A female laboratory assistant at a medical institute: "It's too bad he wasn't killed. In their country [America] they shoot the president, but here all kinds are bred, and they can't shoot them.")

A distinctly positive attitude was noted among about 20 percent of the urbanites questioned (13 percent of the blue-collar workers, 28 percent of the intelligentsia, 25 percent of the Party members, and 32 percent of the administrators. The basic argument here was also materialistic. ("Can you believe it? He had everything, and he gave it up!")

About half of the population have no opinion of their own about Sakharov's activities. It is hardly likely that a similar question about any other political figure in our country would have yielded more positive results. But one should not exaggerate the importance of positive evaluations. They are not "active." As a rule, they are kept to oneself. And with some pressure from the authorities, the great majority of those who gave a positive evaluation to the personality of Andrei Sakharov might come out against him, consoling themselves with the usual: "We're insignificant people. Nothing depends on us."

And yet this passive attitude shows that the proclaimed "unanimity" in no way reflects the genuine mood of the Soviet people, with which someday the Soviet leaders will have to deal.

RAISA LERT

A Man for the People

"I looked around, and my soul was wounded by the sufferings of mankind."

Radishchev wrote that.[1]

The following was written by a hack columnist of the Soviet newspaper *Trud:* "He raised his head from his scientific calculations, looked around, and noticed the general disorder of human affairs."

The columnist was trying to write ironically about Andrei Dmitriyevich Sakharov. But he was so illiterate he didn't realize that he was accidentally paraphrasing the cry of pain uttered by Radishchev, almost the first Russian human rights activist.

Yes, Sakharov raised his head from his scientific calculations. Yes, he looked around. Yes, he noticed "the general disorder of human affairs." But he noticed more than that. He noticed violence, cruelty, injustice, lying, and the violation of human rights. And above all, he noticed the people suffering from all that. And he intervened for them, just as he would have intervened for a child being beaten. His soul "was wounded by the sufferings of mankind."

How he stepped from the gleaming parquet of science onto the sharp, cutting stones of the path of the human rights activist is some-

---

[1]Alexander Radishchev (1749–1802), poet and author of *A Journey from Petersburg to Moscow*. He was exiled by Catherine the Great to Siberia.—Trans.

thing that he himself can no doubt explain better than anyone else. As for the repercussions of his articles, letters, interviews, and essays, that is something known by the whole world (except, of course, for the great majority of inhabitants of the Soviet Union). I can't give as good an account of that as can those who have a more detailed knowledge of his activities for the past ten or fifteen years. That he is a great scientist, even I know, of course. But I am not equipped to discuss his science, physics, any more than I could, say, the Sanskrit language. And if I have decided to write, now, about Sakharov, it is only because I want somehow to convey the feeling of the fascination his personality exerts—something felt by everyone who has chanced to meet him.

I should say that that fascination is radiated, above all, by his *naturalness*. Somehow the word "modesty" doesn't strike me as appropriate when applied to a really imposing personality. It always seems that in the "modesty" of such a person there is a touch of posing: he can't very well be unaware of his own worth. But in Sakharov one finds neither "modesty" nor posing. It's simply that he is what he is, and other people are what they are. All are different, and all are people, and all are interesting in some way, and one feels pain for all of them when they are made to suffer and are deprived of their human rights.

That gift of empathy is perhaps Sakharov's most winning trait, the essence of his charm, the strength of his personality, which influences people more than his articles and books. One may not agree with many things in his books, but it is impossible not to approve of his personality and his deeds. This is not simply goodness, it is a harmonious combination, the cultivation of lofty feelings, of an intellect that grasps everything, and of organic democratism. A rare combination!

There is a memorable line in the Georgian film *Hello, Everybody!*: "It is given to a man to reach the heights not so that he can consider himself superior to others, but so that he can see better." That film is about the Georgian folk artist Nikola Pirosmanishvili, but the words apply equally well to Sakharov. He doesn't measure himself by the height; he sees farther and better.

Many "masters of thought" do not allow one to disagree with them. Often they feel that they possess the truth, by which they are destined to subjugate the world. But one can disagree with Andrei Dmitriyevich. He doesn't preach, he doesn't lay down the law, he doesn't hand out prescriptions. He may make mistakes, and he may acknowledge his mistakes. But when he's certain of something, with his head and with his heart, he is gentle, sweet, and good and stands firm as a rock. (As Martin Luther said: "Here I take my stand; I cannot do otherwise.") That is why he is in exile in Gorky.

He is the only member of the Academy who fearlessly crossed the boundary line of "personal safety measures." He is the only person of that rank who was among those who stood for hours outside the closed doors of "open" trials, trying to support the defendants. He went to visit a prisoner in a labor camp. He made his way on foot along the paths of Yakutia to visit a person in exile. Practical people shrugged their shoulders: others can do that, Sakharov should conserve his time and precious intellect. But he could not do otherwise. He could not spare himself when the destinies of so many human beings were destroyed, hacked to pieces, and mutilated around him. He simply couldn't.

Academician Sakharov is not a social utopian, not an idealistic dreamer, and not a speculative philosopher. He is a scientist—a man with a powerful, sober mind. In one of his interviews, when asked whether he had hopes for victory in the bloodless struggle he was carrying on for freedom and human rights, he said frankly that he had no such hopes. And when asked why, if that was the case, he was carrying on the struggle, he replied (as well as I can remember): "I can't just keep silent and do nothing." This is like Tolstoy's "I cannot keep silent," Radishchev's "My soul is wounded," V. G. Korolenko's declarations, and Emile Zola's *J'Accuse*.

History has known other great hearts and minds who could not ignore human suffering in silence, and all too often they have not been victorious. But they have always influenced other hearts and minds. The declarations of the seven young people on Red Square protesting against the occupation of Czechoslovakia, the organization of the Helsinki groups, and the founding of the samizdat journals were all in the same vein as Sakharov's "I can't keep silent." All that, too, has been almost hacked to pieces and destroyed. The majority of those who "can't keep silent" are in labor camps, in exile, or have emigrated. But whatever the triumph of the department that shall remain nameless, people like Sakharov will remain a radiant example of humanity.

VALERY CHALIDZE

Andrei Sakharov and the Russian Intelligentsia

We must look to the early nineteenth century in order to understand the origins of the Russian intelligentsia. Until then, culture in Russia was closely associated with the court and depended on the patronage of the crown. During the reign of Alexander I—perhaps as a by-product of his unrealized liberal tendencies—some creative talents abandoned the court and, while not yet forming a separate social caste, gathered in intellectual circles which were independent of the state power, and even inclined to oppose it.

The rupture was not abrupt. The poet Alexander Pushkin, known for his love of liberty and for his independence, recognized before his death that the alienation of educated society from authority and from the official state hierarchy was strange and unnatural. Pushkin was partly ready for a reconciliation with the Emperor. But the divorce of culture from the Russian state proceeded, although not everyone recognized the process while it was occurring.

Why have I begun with such ancient history? Because we cannot understand Sakharov's role in Russia today unless we know the epic of the intelligentsia. Sakharov is a heroic figure, and admired everywhere for his appeals on behalf of human rights and fundamental

This speech was delivered at the American Physical Society's Special Symposium to Honor Andrei D. Sakharov, New York, January 26, 1981.

freedoms. But he is also a tragic actor in the drama of Russian history and of the Russian intelligentsia.

Who is guilty in the conflict between state and intelligentsia in Russia? Both parties, I believe. The authorities because they want to subordinate culture to their own purposes and turn the intelligentsia into propagandists for imperial greatness. The intelligentsia because in warring against the state for two centuries, they turned their backs on normal politics and have played an exclusively negative role, first as critics of the prevailing order and then as destroyers of that order. The Russian intelligentsia constitute a unique social group. I have not discovered any western counterpart. Intellectuals do exist in the West, and they may oppose the government, but they do not form a separate, coherent caste.

The intelligentsia fostered the fall of the Russian Empire, but after the Revolution, the intelligentsia remained in opposition and the new regime made war upon the social group which had created it. The very nature and beliefs of the intelligentsia seem to require opposition to the government in power. We have learned much about the bloody persecutions of the intelligentsia during the Soviet era, but we know too that its traditions survived and continue to enrich world culture. But the tragedy for the Russian nation is that the intelligentsia have excluded themselves from practical politics. The intelligentsia as a matter of principle refuse to participate in the decision-making process of the state. Moreover, the intelligentsia scorn the common man's striving for a career and success. I remember my own friends in Russia—they thought the word "career" somehow disgusting. Their ideal is a selfless dedication to culture, to the ideals of art and science with almost no regard for recognition by society.

From one perspective, Sakharov's way of thinking, his ethical principles, even his opinions, are typical of the Russian intelligentsia. On the other hand, he not only served in official posts; he rose high in the state hierarchy and did much to make strong the existing regime. It is well known that Sakharov made notable contributions to the military applications of thermonuclear reactions. For many years he occupied a leading post in the military-industrial complex, and he received the highest state awards for his accomplishments. Similar cases were known earlier. Intellectuals had served successfully in official posts, but as a rule they were then no longer considered intelligentsia in the Russian sense.

But Sakharov never stopped being an *intelligent*. His faculty for independent thought about society as well as about science never atrophied, although social criticism is a taboo subject for private Soviet citizens. Sakharov made no effort to camouflage his opinions. After working inside the establishment for many years, in 1968 he published his outspoken views on the course of Soviet society and on the dangers threatening the world. His essay *Progress, Coexistence, and Intellectual Freedom* was a sincere attempt to initiate a dialogue to which the regime could well have responded. His act was, however, too unexpected, and the state hierarchy expelled Sakharov. The regime once more displayed the symmetry of its relations with the intelligentsia. The intelligentsia totally reject the regime, and the regime replies in kind. Of course, the symmetry is not complete, because the authorities send the intellectuals to prison and exile.

Sakharov's public activity has turned him into a symbol of liberty, of opposition to tyranny. He has said many wise and good things. But I believe that he will be remembered in future history books as one of those rare intellectuals who dared to break down the wall between power and culture, disregarding the moral taboos of his social group and the lack of understanding of the authorities.

Earlier attempts to build bridges between state and society foundered because they were out of phase. Alexander II introduced important reforms which should have been welcomed by the intelligentsia, but they disdained gradual progress. In the second half of the nineteenth century, they began sharpening the ax which cut off many heads of the intelligentsia after 1917.

History is ironic. The intelligentsia might have gone to meet Alexander II's reforms or Nicholas II's creation of a parliament. But the moments were lost and instead the current dissident movement had to begin by seeking a dialogue with Mr. Brezhnev. By then the authorities were deaf and blind. They rejected the intelligentsia's attempt, and so the country remains *at war with itself*.

Forecasts are risky, but I believe it possible and desirable that in the future the situation in Russia will lead the regime to undertake cooperation with the intelligentsia without insisting that they abandon their principles or tell lies. And perhaps circumstances will change sufficiently so that in Russia as in other civilized countries the intelligentsia will no longer consider participation in the political process shameful and will be willing to assist in the government of their country. This may sound utopian, but Sakharov's example shows that it is possible. He worked within the government establishment and still remained an honest man and a member of the Russian intelligentsia.

Sakharov's tragedy is not his interrupted career, not the lack of success of many of his human rights initiatives, not even his current exile in Gorky. His tragedy lies in his attempt to overcome the two-hundred-year antagonism between the intelligentsia representing the culture of society, and the authorities representing the power of society.

No single cause can explain the tragic course of Russian history for the last two centuries, but the hostility, the lack of mutual understanding, between the state and the intelligentsia was surely a significant factor. Elimination of this conflict is absolutely necessary for Russia's health in the future. The ice must be broken by brave individuals. Sakharov was the first and most prominent man of our age who—while holding important state posts—remained a member of the intelligentsia and openly expressed their beliefs.

BORIS ALTSHULER

ON SAKHAROV

I met Andrei Dmitriyevich in 1968 when he agreed to serve as opponent when I defended my doctoral dissertation on the general theory of relativity. At the time I was twenty-nine.

In August 1969 we were on the same plane going to an international conference on gravitation in Tbilisi. Because of a storm over the main ridge of the Caucasus Mountains, the plane didn't fly on to Tbilisi. We spent the night in chairs at the airport in Mineralnye Vody. That was a long time ago—on my time scale, at any rate. I had defended my dissertation at that same Physical Institute of the USSR Academy of Sciences (FIAN) where Sakharov began to work after he had been barred from secret activities. From that time on, almost every Tuesday we met at the Tamm theoretical seminar. (Igor Evgenevich Tamm had died in 1971, but the name of the seminar was preserved.) And now for more than a year we haven't met. I can't reconcile myself to that.

In these notes I can't pretend to completeness and consistency in setting forth the facts of Sakharov's biography, so I shall confine myself to a few subjective remarks.

Sakharov was a student in the physics department of Moscow State University when the Great Patriotic War began. By government decree, his whole class was evacuated to Ashkhabad in the fall of 1941. They completed their course of study under an accelerated program, and in 1942 the young engineers went to work for the Ministry of Defense. At an ammunition plant in the city of Ulyanovsk, Sakharov became the author of several inventions in methods of quality control which were promptly introduced in production during the war.

After the war Sakharov did graduate work under Igor Tamm. In 1948 he was included in a scientific research group assigned to carry out an especially important government task. (After a scientific seminar, Tamm asked Sakharov and another young theoretical physicist to stay behind, and said to them: "Get ready, we'll soon be going away." "Where and why?" "I don't know myself," answered Tamm.)

The explosion of the first Soviet atomic bomb occurred on August 29, 1949, and on August 12, 1953, a hydrogen bomb was exploded. I. N. Golovin's book *I. V. Kurchatov* (Moscow: Atomizdat, 1967) recounts something about this period. The author says: "Sakharov prompted us to the solution of a second, equally great atomic problem of the twentieth century: obtaining inexhaustible energy by means of burning ocean water!"

Golovin's book very accurately conveys the state of uplift, of joyous excitement, characteristic of that period (which we now know to have been a very frightful one). I remember that "inspiration"; and I remember the general grief (including my own) and the feeling of being lost when Stalin died. Almost everyone in the country, to one degree or another, felt the action of that ideological narcotic. (According to the testimony of eyewitnesses, in the concentration camps people thought differently.) But Sakharov showed a certain independence of thought even then. It was precisely during this period that he repeatedly rejected suggestions that he join the Party. (Scientists of the special group were "forgiven" what others could not be forgiven.)

The gradual, tortuous process of awakening began after 1953. What distinguished Sakharov from many others was the fact that for him there never existed any distance between conviction and action, between words and the main strategy of life. Every routine test, by raising the general level of radiation in the atmosphere of the earth, entails thousands of unknown victims in the long run. According to the testimony of Sakharov himself, it was because of these considerations that he began to speak out for banning tests. In a country where people "were not counted," a concern for those obscure people was rather "strange," not to mention the specific actions engendered by that idea. But Sakharov felt personal responsibility for the tragedy of those people. As a result he managed to facilitate a conclusion of the Moscow treaty on banning nuclear testing in the three environments.[1] In 1962, talks with the U.S. on banning tests had long since bogged down because of the controversial question of monitoring underground nuclear explosions. Sakharov went to the Minister of Medium Machine Building, E. P. Slavsky, with the idea of excluding that controversial "fourth" environment from the draft of the treaty on the grounds that such a Soviet initiative would improve the USSR's position at the UN. Slavsky passed the idea on to the then Soviet representative at the UN, Yakov Malik. The further route of the idea is unknown; but judging from the results, it pleased Khrushchev, who, as we know, attributed great importance to improving the Soviet position at the UN. Thus in the summer of 1963 the Moscow treaty arose.

The thesis of "compassion for people" lies at the basis of all Sakharov's social actions. When the scale of the mass murders of past years became known, he experienced it as a personal drama. Such a thing must not be repeated. Sakharov has never felt himself to be a "little man" who knows that "you can't change anything anyway," and he has completely taken upon himself the responsibility for what happens. There are situations in which one cannot be passive. Passivity is also a kind of action, and sometimes a very dangerous one. For Andrei Dmitriyevich, insofar as I can judge after many years of acquaintanceship, such is his inner position—a part of his personality.

For some reason it turns out that the ideas set forth by Sakharov usually acquire fundamental importance.

In science: the magnetic containment of plasma to obtain a controlled thermonuclear reaction; the idea of the instability of the proton to explain the baryon asymmetry of the universe. And there are other ideas whose significance, quite possibly, will emerge in the future. (Sakharov's scientific activity is a special subject, and I shall not dwell upon it here, although it was precisely at scientific seminars

[1]The atmosphere, outer space, and the ocean.—Trans.

that I saw him regularly. I can testify, however, that he worked at science at all times and in spite of everything.)

In the social sphere: In his "Thoughts . . . ," published in 1968, Sakharov put forth the idea of the danger of any total ideology—social, nationalist, great-power, etc. In his Nobel lecture he formulated with great clarity the conception of the priority that must be given to the necessity of observing human rights, and the profound connection between that problem and the problem of preserving peace on earth. Today these ideas of Sakharov's are accepted by everyone, from the Eurocommunists to the conservatives, and perhaps they have even given a slight nudge to certain Soviet ideological stereotypes. Of course that did not happen right away and of itself.

Herewith a few landmarks. (The selection is perhaps subjective and incomplete. Also, I realize the complexity of such a phenomenon as the historical process, and in no way do I want to minimize the heroic efforts of other people.) The interview of August 21, 1973, in which Sakharov, taking full responsibility for each word he uttered, said true détente was impossible without Soviet society's becoming more open. By way of a step in that direction, his support of the famous Jackson Amendment in the U.S. Congress, which created a real stimulus for observing one of the basic human rights—the right of free choice of one's country of residence. The book *My Country and the World* (1975), in which much of importance was said, and which fortunately was read in the West (although, unfortunately, it was not read in the USSR).

Igor Tamm gave a fine description of Sakharov as a scientist (see "Physicists on Sakharov," later in this collection). I might add that those traits of Andrei Dmitriyevich noted by Tamm are manifested in all his activities. He is able to find the "sore points" in a problem, go beyond the framework of the existing, and create something new.

The publication of "Thoughts . . . " in the West and Sakharov's subsequent actions made his name a legend and were perceived as madness by many people in the USSR. But not by all. In the spring of 1970 Igor Tamm asked Sakharov to represent him at the ceremonies, held at the university, in connection with the award of the Lomonosov Medal to Tamm, and to read his laureate's lecture. (Tamm himself was very ill at the time, and could not leave his home.) That expression of trust was very meaningful for Andrei Dmitriyevich, since he was already an "untouchable."

In 1969 Sakharov gave almost all of his savings (134,000 rubles) to the Red Cross for the construction of a cancer hospital in Moscow. (The Red Cross officially thanked him, whereas the director of the cancer center, Academician N. Blokhin, manifested no such politeness. Eleven years later, in 1980, at an international conference of scientists in West Germany, Blokhin said a number of unpleasant things about Sakharov, by then exiled from Moscow.)

In 1972, having become convinced that his suggestions and appeals to the government went no farther than the archives of the KGB (just as, today, petitions from the West in defense of Sakharov come to rest with the KGB), Sakharov, who until shortly before had been supersecret, did the impossible. In the autumn of that year he and his wife, Elena, received foreign correspondents at their home. At the time, Moscow physicists were saying: "The existence of Sakharov and Solzhenitsyn violates the law of the conservation of energy." I would like to believe that that judgment is wrong, and that the law of the conservation of energy is more fundamental than the law of the conservation of fear.

JULY 1978. The last day of Anatoly Shcharansky's trial. In a little side street in the center of Moscow, a crowd of people. Among them are Andrei Dmitriyevich and his wife. They are tired, just having returned from Alexander Ginzburg's trial in Kaluga. No one is allowed even into the courtyard of the courthouse building. Many agents in civilian clothes. At a barrier guarded by policemen, an elderly woman—Shcharansky's mother. We stand there a very long time. Our own powerlessness is agonizing, and demands some kind of respite, but nothing can be done. The trial will soon be over, and we'll break up. Then Sakharov's voice is heard: "Let his mother in! At least for the reading of the sentence!" The crowd bunches against the barrier. Andrei Dmitriyevich utters the only possible and necessary words in that tragic situation. Then he walks off, pale, and takes some kind of pill for his heart trouble. (Shcharansky's trial was an attempt to shift the struggle against dissidents and Jews into the usual channel of spy mania. At the time, it was possible to check that dangerous tendency. But Anatoly was convicted and his sentence isn't up until 1990. It's frightful to look at those figures.)

There have been very many such episodes in recent years. The idea that in the struggle for human rights the main thing is to help particular people is something that Sakharov never tires of demonstrating, with a kind of pedagogical insistence. Essentially, such is the moral foundation of the whole human rights movement. Tatyana Osipova, who was tried in Moscow in early April of 1981, said in her final plea: "I consider the defense of human rights my life's work, because the violation of those rights leads to human tragedies." She was sentenced to five years in a penal camp and five years of exile. Such sentences are devastating, especially when one realizes that neither Osipova nor any of the other human rights activists has ever called for violence, or resorted to it, and that their only weapon, as they see it, is publicity.

There are neither more important people nor less important people; the life of every person contains infinity; the universe is not measured quantitatively. These ideas were not conceived by Sakharov, but they are extraordinarily organic for him. His Nobel lecture and many of his statements contain very long lists of repressed persons. It is agonizingly hard for him to stop listing names, and he does so with a feeling of guilt toward those he hasn't identified. To some people that might seem tiresome. Two years ago the Voice of America, in broadcasting one of Sakharov's statements, mentioned only the first few names on his list, replacing the rest with "and others." That provoked a sharp protest from Andrei Dmitriyevich—a deep personal indignation. If this position of Sakharov's were accepted broadly enough, perhaps humanity would be saved. And conversely, a disregard for the principle of the absolute value of each human life is fraught with millions of human corpses, as history has shown.

For a long time, the activity of the human rights movement in the USSR (including that of Sakharov) obstructed attempts to weaken the restrictions imposed on the KGB after the death of Stalin. That has often been confirmed in experience. Just what the mechanisms of that influence are is hard to say. The machinery of power in the USSR is a "black box," and one probably shouldn't rack one's brains over its puzzles. "What will happen?" "Where is everything headed for?" Here, everybody asks everybody else these questions. But the answer may depend on one's behavior at the given moment. Sakharov has repeatedly shown the effectiveness of his own actions.

Recently, something has shifted in the "black box," and the KGB has more freedom of action than before. This has been manifested in the exile of Sakharov and the broad repressions of human rights activists, including women, involving a special element of harshness (Tatyana Velikanova, Tatyana Osipova, Irina Grivnina, Malva Landa, Oksana Meshko, Olga Matusevich). Also, in the provocative arrest of the mathematician Victor Brailovsky, secretary of an unaffiliated scientific seminar and a Jewish "refusenik," in precautionary repressions (Anatoly Marchenko, Genrikh Altunyan), and in the jamming of radio broadcasts. What will happen? "The important thing is what has already happened," Andrei Dmitriyevich answered when I put that question to him some four years ago, after the arrests of Orlov, Ginzburg, and Shcharansky. That answer is still the right one, and probably always will be. The liberation of the American hostages was the result of certain fundamental efforts. What will happen to Sakharov, Orlov, and other human rights activists may depend upon the efforts of world opinion. It is essential to find ways of talking directly with the top echelon of power in the USSR—i.e., with the only ones fully empowered to "make decisions"—and that is not simple. (Academician Sakharov has talked, and still talks, about the importance of this latter circumstance.)

"What will happen" may depend upon the position of Soviet public opinion. The position of the USSR Academy of Sciences regarding the exile of Sakharov is well known: silence and passivity; that unprincipled passivity which is dangerous. Not a single academician has demanded the most elementary thing: to let Sakharov speak—to listen to him at a session of the Academy. It is not to be ruled out that an appeal to Brezhnev by several Soviet academicians might return Sakharov to Moscow.

As it happens, I saw Sakharov at a 1973 seminar held at the FIAN the day after his apartment had been visited by people calling themselves members of the Black September organization. He talked about it as a kind of vexing, absurd event. "Basically, Sakharov can't be frightened," his wife once said. "You can kill him, but you can't make him disown what he thinks." Andrei Dmitriyevich is constantly thinking. As I understand it, he always feels the distance between fundamental circumstances and accidental, superficial ones like, for example, the visit of the "terrorists." He can't be frightened; but one can make things very painful for him if one uses his dear ones—his wife, his children—to that end. Today Liza Alekseyeva, the fiancée of his son, has become such an object. "The very fact of holding hostage someone associated with me is completely unbearable for me." (Letter to A. P. Alexsandrov, October 20, 1980.) These words of Sakharov's should be taken with the utmost seriousness.

One of the main factors forming the character of people in this country is the lack of information. People here are very poorly informed, *inter alia* about Sakharov. The distribution of samizdat is plainly inadequate for the development of that "collective effect" which is called social consciousness. Under these conditions, various kinds of psychological "grimaces" are encountered. Many people are irritated by Sakharov's activism. Such a reaction is like the remarks one hears from narrow-minded people: "The lazy Poles go on strike because they don't want to work. We live worse than they do, but we don't strike. So what are our boys waiting for?" (Needless to say, there are other opinions, but that stereotype is very widespread.) Yet I daresay that social psychology is changing—very rapidly on the historical scale of time—under the influence of conditions. Today extreme conditions of the isolation-chamber type exist. But if there will be information (if only from outer space, on television screens—something Sakharov has repeatedly called for), within ten or twenty years (or perhaps even earlier) the Polish word "solidarity" will appear in the Russian language.

On that optimistic note I come to a conclusion. What should one wish dear Andrei Dmitriyevich on his birthday?

That he, despite everything, may for many long years preserve his working capacity and clarity of thought, necessary for understanding and giving form to the world.

Health to him and Elena, which in the present unthinkable situation is especially urgent.

That Liza Alekseyeva receive permission to leave the USSR, and that she, Alyosha, and all the Sakharovs' dear ones be able to return when they want to. That all people be able to cross national borders when they want to, in any direction.

That Sakharov's Soviet colleagues unbend.

That his foreign colleagues, whose help up to this time has been so substantial, speak out in his defense, as in the defense of other prisoners of conscience, with a tenfold "Sakharovian" vigor.

That he return to Moscow, and that all his friends be released from prisons and labor camps.

That all political prisoners in the USSR be amnestied; that legality and law and order reign in the country; and that publicity and criticism be unpunishable norms of life.

That mankind master thermonuclear energy and avoid thermonuclear catastrophe.

But that is something to be wished for all of us.

II

GRIGORII S. PODYAPOLSKY

My Conversation with the Director of the Institute of Geophysics of the USSR Academy of Sciences

The transcript printed below is a document from a very recent period. Its author, Grigorii S. Podyapolsky, a research associate at the Institute of Geophysics, made the transcript while (one might well say) "there were still fresh tracks" left from an administrative rebuke delivered by the institute's director to a group of mathematicians who had spoken out in defense of one of their colleagues who had been subjected to psychiatric persecution.

Later, Podyapolsky was a human rights activist, a member of the Initiative Group for the defense of Human Rights in the USSR, and a member of the Human Rights Committee. He died prematurely in 1976 at the age of forty-nine while on an official trip—a trip he had been sent on "during the congress."

The conversation quoted below took place under the following circumstances. In the spring of 1968, along with a number of other associates of our institute and some mathematicians from other Soviet scientific institutes, I signed a letter in defense of the mathematician Esenin-Volpin, who had been forcibly confined in a psychiatric hospital for reasons having nothing to do with the state of his health or with medicine in general. That letter, the original text of which I unfortunately do not possess,[1] was very short, consisting of only three sentences. The first stated the fact of forcible confinement in a hospital. The second pointed out that the confinement in a psychiatric hospital of a quite able-bodied and talented mathematician would unquestionably have a damaging effect on his health and mental state. The third contained a request that Esenin-Volpin be returned to normal conditions of life and work.

The letter was sent to three addressees: the Minister of Public Health, the chief psychiatrist of the city of Moscow, and the procurator (either of Moscow or of Moscow Oblast, I don't remember which). We received no answer from any of these officials. But in some manner unknown to me, the letter suddenly showed up in several very awkward places, most notably the KGB, the presidium of the USSR Academy of Sciences, and abroad, from where it was broadcast in Russian by several foreign radio stations.

The last-named circumstance served as the pretext for an "educational campaign" for the signers of the letter. One of the phases of that campaign that involved me personally was the conversation presented below. It was not the first phase, nor the last. Before it, attempts had been made to get the "signers" in a small number of sections to repent. (Later, those attempts were recognized as "inadequate.") After it, a rebuke was administered at a general meeting of the department of the institute.

The conversation itself took place under the following circumstances. All of the "signers" were notified in advance that on such-

[1]The letter was subsequently found. Here it is.
To: The USSR Minister of Public Health, Petrovsky
    The Chief Psychiatrist of the City of Moscow [Enushevsky]
    The Procurator of the City of Moscow
    It has come to our attention that the outstanding Soviet mathematician and well-known specialist in the field of mathematical logic A. S. Esenin-Volpin was forcibly confined in a psychiatric hospital at Stolbovaya Station (100 kilometers from Moscow) without prior medical examination, and without the knowledge or agreement of his family. The forcible confinement of this talented and quite able-bodied mathematician in a hospital for seriously ill mental patients, and the conditions under which he now finds himself, are badly traumatizing to his psyche, damaging to his health, and belittling of his human dignity. We ask you to intervene promptly and take steps to ensure that our colleague can work under normal conditions.
                                            [Signed by ninety-nine mathematicians]

and-such a day, at such-and-such an hour, they would be having a talk with M. A. Sadovsky, the director of the institute. Therefore, all of them were supposed to be at their working places at the given time. One by one, the director's secretary called them on the telephone and summoned them to the director's office, so that in each case the talk was a tête-à-tête.

I did not make this transcript at the time of my own talk with the director, since I didn't feel that the conversation merited it. By comparison with other measures of the same kind carried out at the same time at other academic and nonacademic institutions, it seemed quite ordinary and relatively decorous. But now, after about a year, I have for various reasons changed my views as to the importance of that conversation, and I felt that it is worth bringing to the attention of the public.

In setting forth the conversation I have tried to be as objective as possible, to avoid any exaggeration, and not to omit any substantive detail. I cannot vouch for the textual accuracy of individual sentences or their correct sequence, and I may have omitted certain nonsubstantive things that were said. But I can vouch for the accurate conveyance of the basic tenor of the conversation, for a faithful reflection of its tone, for the absence of any substantive omissions, and for the total absence of any invention.

I (entering): Hello, Mikhail Aleksandrovich.

M. A. SADOVSKY (in a cheerful tone inviting frankness): Hello. Well, do you admit that you were an instigator?

I (imitating his tone): Why, how can you say that, Mikhail Aleksandrovich?

[Note: This was the gospel truth; I had not been an instigator. But Sadovsky obviously didn't believe me. I don't know whether it was because of his self-assurance, or because of nobler motives, that he didn't ask me the question I expected: "Then who was it?" But he didn't, and the question of the instigator was dropped.]

SADOVSKY (after asking me to sit down, in a completely different voice, monotonous and totally expressionless—as it turned out later he had repeated this tirade to all the "signers," and I was one of the last to be called): In the year nineteen hundred and something-or-other, when the late President Kennedy was still alive, the U.S. Congress allocated to the Central Intelligence Agency of the United States of America the sum of one million dollars to carry out subversive activities against the Soviet Union and other countries of the socialist camp. That money was used to set up provocations, to recruit agents and send them here, and to carry on propaganda. . . . [I don't remember, verbatim, what he said from that point on. But the gist of it was that those of us who had signed the letter had been lacking in vigilance, and that international imperialism had led us around by the nose.] Do you understand that?

I: No, I don't understand.

SADOVSKY (wearily): But I just explained everything to you. . . .

I: But after all, Mikhail Aleksandrovich, you and I are scientists—

SADOVSKY: Come, now! No demagogy!

[*I don't know why he was offended by my saying that "you and I are scientists, after all." In any case, that was his only outburst of anger. The tone of all the rest of the conversation was irreproachably correct.*]

I: The thing is, I want to say that I don't see any connection between the dollars of the CIA and a letter sent in defense of a man subjected to a very trying and unfounded action—a letter addressed not just anywhere but to Soviet state organs. The initiative for the letter arose among Soviet mathematicians who have a high opinion of the talented scientist Esenin-Volpin, and who were disturbed by what might happen to him. It certainly did not arise among agents of the imperialist intelligence services.

SADOVSKY: But after all, the letter ended up abroad.

I: Yes, it ended up abroad and in other places.

SADOVSKY: Don't you think that the fact that the letter ended up abroad was a provocation?

I: Yes, it's very possible that it was a provocation.

[*Note: On this point, we were obviously speaking different languages. Sadovsky, of course, meant that it was a provocation by the CIA, while I meant that it was a provocation by different organs with a completely different aim. Whether Sadovsky understood that difference, I don't know. I believe that he did; but he pretended that he didn't.*]

SADOVSKY: Doesn't it seem to you that you should have foreseen such a possibility and thought long and hard before signing such a letter? After all, if you'd been signing a document involving money, you'd have thought about it. Doesn't it seem to you that you bear personal responsibility for the fact that that the letter ended up abroad?

I: No. I can answer for the letter so long as it was in my hands. But once it was sent to the addressees, I can bear no responsibility for what happened to it.

SADOVSKY: But the fact remains that many mathematicians turned out to be different from you. They were asked to sign the letter, and they refused. Why do you think they did?

I: I think they were simply afraid.

SADOVSKY (ironically): So you regard yourself as very brave?

I: I don't regard myself as very brave. It seems to me that I merely behaved in a way natural to a normal person. But a few people were afraid for some reason.

SADOVSKY: So you are braver than Academician Kolmogorov and Academician Aleksandrov? They didn't sign the letter.

I: It would have been illogical for them to sign it. They had already signed another letter, from the two of them, on the same subject. The collective letter was written because there had been no reaction to the letter from Kolmogorov and Aleksandrov.

SADOVSKY: Don't you know how the imperialist intelligence system works? Do you know that on that same evening when Esenin-Volpin's mother asked Kolmogorov and Aleksandrov to write a letter, somebody called one of them from Paris asking for information about that letter?

I: No, I didn't know about that. But if that is so, I don't see at all what damage could have been done to the Soviet Union by our letter. Such damage could have consisted only in the fact that word about Esenin-Volpin's confinement in a psychiatric hospital had been leaked abroad. But from your statement it is evident that it had leaked abroad even before our letter so much as existed in rough draft.

SADOVSKY (to my amazement, this simple bit of logic confused him for a moment): But after all, the letter was broadcast by the Voice of America and was used for propaganda against the Soviet Union.

I: Keldysh's[2] speech at the plenum of the Moscow Committee was also broadcast by the Voice of America, and was used for propaganda against the Soviet Union. Do you consider Keldysh responsible for that?

SADOVSKY: But Keldysh's speech was broadcast in excerpts, while the full text of your letter was read.

I: Keldysh's speech was long, but our letter consisted of only three sentences. There simply wasn't anything to be condensed.

SADOVSKY: Well, it's plain to see I won't succeed in convincing you.

I: Obviously, you won't.

SADOVSKY: Well, then. Goodbye.

I: Goodbye. (Turning at the door) Once again I can assure you that when I signed the letter I was in my right mind, and my memory was sound.

As is evident from the text of the transcript, the conversation was not so much sharply animated as listless. Both the prosecution and the defense performed rather weakly, and neither utilized many obvious resources. Thus, I was not asked the question that later figured at the general meeting: "Why don't you believe the Soviet organs of public health, which found Esenin-Volpin insane?" To the credit of Sadovsky, I must note that none of his questions was of the type

[2]Mstislav Keldysh, then president of the USSR Academy of Sciences.

asked at an interrogation, such as "Who gave you the letter to sign?" Or "Where did you learn about the letter written by Kolmogorov and Aleksandrov?" Or "Do you know Esenin-Volpin personally?" Nor did he ask me: "Would you have signed the letter if you had known that it would end up abroad?" My impression (and I may be mistaken) was that Sadovsky was carrying out—with evident reluctance and no enthusiasm—a function imposed upon him. But I must admit that in the course of that conversation I also failed to bring into play a basic argument. I mean the basic moral argument of the defense: that in the final analysis the responsibility for any consequences rested with the initiators of the action against Esenin-Volpin, and not with anyone else.

Thus, the conversation could not have convinced anyone of anything and in my opinion it was not aimed at doing so. But for future generations it may be of some interest because of its preposterousness and absurdity, which are typical of present-day "educational measures."

June 14, 1969

Physicists on Sakharov

Academician I. E. Tamm

He [Sakharov] has a splendid attribute. He takes a fresh approach to any phenomenon, even if it has been researched twenty times and its nature has been established twenty times. Sakharov deals with everything as if he had before him a blank sheet of paper, and owing to this he makes astonishing discoveries.[1]

M. A. Leontovich and B. B. Kadomtsev

In 1950 a much more important, new line of approach in plasma physics developed, which also has its roots in atomic physics. I. E. Tamm and A. D. Sakharov introduced the idea of magnetic thermal isolation of a plasma for producing a controlled thermonuclear reaction. Preliminary calculations showed that it is technically possible to achieve the conditions for a self-sustaining reaction in a toroidal apparatus.[2]

This selection, and the succeeding articles on Sakharov's scientific contributions—"Controlled Thermonuclear Reactions"/"Magnetoplosive Generators," "Sakharov's Papers on Fundamental Problems of Physics," "Sakharov's Problem Games," and "Sakharov's Textbooks and Popular Science Articles"—have all been translated by Wendell Furry.

[1]From M. Romm, *Clarity of Vision*, Ekran, 1964, Moscow, Iskusstvo, 1965, p. 133.
[2]Academician M. A. Leontovich and Corresponding Member B. B. Kadomtsev, Acad-

Academician A. P. Aleksandrov

As a third example[3] of our science's possibilities we may take the development of work in plasma physics for the purpose of realizing a thermonuclear reaction under quiet, controlled conditions. This problem is especially important owing to the fact that there are enormous natural supplies of deuterium, which offers the promise of relieving mankind, practically forever, from concern about energy supplies.

I. E. Tamm and A. D. Sakharov suggested the idea that thermal isolation of a hot deuterium plasma could be achieved by means of a magnetic field. The charged particles making up the plasma cannot move across a magnetic field. Making use of this, one could devise a configuration of magnetic fields surrounding a hot plasma on all sides, so that the plasma could not leave the designated space.

In the last years of his life, I. V. Kurchatov devoted himself with extraordinary energy to organizing work on this important approach. In addition to the Plasma Research Section of the Atomic Energy Institute, in which, under the direction of L. A. Artsimovich, work on thermonuclear synthesis was a major activity assisted by a strong group of theorists led by M. A. Leontovich, I. V. Kurchatov organized work on synthesis in other sections of the institute, directed by E. K. Zaboskii and I. N. Golovin. Included also in the work were large groups in the Ukrainian Engineering Physics Institute, the Leningrad and Sukhumi Engineering Physics Institutes, and the Lebedev Physical Institute of the Academy of Sciences of the USSR.

The Leningrad Electrophysical Institute was brought into the work of constructing large thermonuclear installations. I. V. Kurchatov showed initiative in arranging international collaboration in this field, and its development is progressing.[4]

---

emy of Sciences of the USSR, "Plasma Physics," in collection *October and Scientific Progress*, Moscow, 1967, Vol. 1, p. 240.

[3]The first example was the development in the USSR of technology for high-energy physics; the second was the construction of experimental reactors.

[4]"Nuclear Physics and the Development of Atomic Technology in the USSR," in the collection *October and Scientific Progress*, Moscow, 1967, Vol. 1, pp. 209–10.

.   .   .

Editorial note: In our time the importance of the research done in
the field of controlled thermonuclear reactions by Tamm and Sakharov, and its foundation-laying character, are becoming more and
more obvious. And yet, no mention of or reference to their work is
allowed; the mere appearance of Sakharov's name in a positive context is altogether impermissible, even in Soviet scientific literature.
And indeed, an intermediate usage is employed—articles appear
without the authors' names, as if they were perhaps produced by the
entire Soviet people.

> The construction of the apparatus was based on the develop-
> ment of ideas that were formulated in 1951 in Refs. 2 and 3.
> In these papers, which initiated the work on controlled ther-
> monuclear reactions in the Soviet Union, the project of a mag-
> netic thermonuclear reactor was proposed. . . .
> 2. "Theory of a Magnetic Thermonuclear Reactor" (Part 1).
>   Proceedings of the Geneva Conference, Vol. 1, pp. 3–
>   19.
> 3. (Part 2) ibid., pp. 20–30.[5]

Part 1 was originally published with the byline I. Tamm; Part 2 with
the byline A. Sakharov.

Is not the ascription to these papers of an anonymous, folklorelike
character the highest form of acknowledgment of the services of Andrei Dmitriyevich Sakharov in Soviet science?

Academician I. E. Tamm

In the field of controllable thermonuclear reactions Andrei Sakharov
not only proposed the fundamental idea of the method, on the basis
of which one can hope to realize such reactions, but also carried out

[5]"Progress of Science and Technology." Series: *Plasma Physics*, Vol. 1, Part 1, edited
by V. D. Shafranov. Moscow, 1980. Tokamaki "Tokamaks," V. S. Mukhovatov.

extensive theoretical researches on the properties of high temperature plasma, its instabilities, and so on. This assured the success of the experimental and technical studies, which have won general worldwide recognition.[6]

[6]"Theoretical Physics," in *Science and Life*, 1967, No. 10, pp. 110–15.

BORIS ALTSHULER

Controlled Thermonuclear Reactions *and* Magnetoplosive Generators

1. Controlled Thermonuclear Reactions

In 1950, Andrei Sakharov, together with I. E. Tamm, advanced an idea which is probably his main scientific and inventive achievement. This is the proposal to realize a controlled thermonuclear reaction for purposes of energy production by using the principle of magnetic thermoisolation of a plasma (see articles on Sakharov and Tamm in the *Great Soviet Encyclopedia*). A controlled thermonuclear reaction, like the reaction occurring in a hydrogen bomb, consists of fusion of the nuclei of isotopes of hydrogen, deuterium, and tritium, with formation of helium nuclei and release of energy—not in an explosion, however, but under conditions of an industrial installation, a thermonuclear reactor. Unlike the chain reaction of the fission of nuclei of uranium and plutonium in atomic bombs or in the reactors of atomic power stations, a thermonuclear reaction is possible only at temperatures of tens or even hundreds of millions of degrees.

Sakharov and Tamm showed that when charged particles, nucleons and electrons, move in a magnetic field of a particular configuration the loss of heat is so greatly reduced that it becomes possible in principle to heat the plasma to the necessary temperature and to maintain this temperature for a time sufficient for a thermonuclear reaction to occur. I. V. Kurchatov reported on their work on April

25, 1956, in a remarkable lecture at the English atomic center in Harwell during a visit to England with Khrushchev and Bulganin. It was also published in the proceedings of the Geneva Conference on the Peaceful Uses of Atomic Energy, and in a collection under the general title "Theory of a Magnetic Thermonuclear Reactor" (MTR). Parts 1 and 3 are articles by Tamm, and part 2 is an article by Sakharov. In a brief introduction Sakharov writes:

> The paper by I. E. Tamm explains the properties of a high-energy plasma in a magnetic field which offer the hope of realizability of an MTR. Here we shall examine other questions in the theory of MTR, namely: (1) Thermonuclear reactions. Bremsstrahlung. (2) Calculation of a large model. The critical radius. Edge effects. (3) The magnetization power. Optimal construction. Productivity in active substances. (4) Drift in a nonuniform magnetic field. Suspended current. Inductive stabilization. (5) The problem of plasma instability.

Sakharov and Tamm are recognized as the pioneers in this work. Further research was continued under the direction of L. A. Artsimovich. Here is what was written about this in the USSR about ten years later (in subsequent editions the name Sakharov is absent): "Kurchatov told his English audience about a most original idea put forward in 1950 by Tamm and Sakharov—that of thermally isolating the plasma by the use of a magnetic field" (from P. A. Astashenkov's biography of I. V. Kurchatov, Moscow, 1967).

And here is a passage from I. N. Golovin's biography of Kurchatov (Moscow, editions in 1967 and 1972, pp. 81–82). In a conversation between Kurchatov and his associate director (whose name is not given) on New Year's Eve, December 31, 1950:

> Associate: "Igor Vasilevich! MTR is the greatest problem in the release of nuclear energy. You have successfully solved the first problem. No one still doubts that an electric station using nuclear fission will work. Sakharov has brought us up to the solution of the second, no less magnificent atomic problem of the twentieth century—the production of inexhaustible energy

by burning water from the oceans! This is a problem to which he did not hesitate to dedicate his whole life."

Kurchatov stopped. A beaming smile lit up his face. "You are excited, young man! You say—a great problem! . . . " His face grew serious. "Yes . . . a great problem . . . a problem for humanity." Kurchatov again paced about, stroking his beard. "The greatest problem! And how will you produce the hot plasma?"

"It's not clear."

"By no means clear! Yes, young man, by no means clear."

"But that is the basic problem."

Kurchatov began, with his usual urgent persistence, to discuss in detail how to make the plasma by an inductive method, fitting the toroidal vacuum chamber with an iron core and a primary winding. . . . In only a few months' time a laboratory staffed by a hundred scientists was in operation, founded by Kurchatov and directed by Artsimovich. M. A. Leontovich headed the theoretical research.

One of the results of years of work by the large group of Soviet scientists was a system known as the tokamak. This system was very close to the original ideas of Sakharov and Tamm, who had considered in particular a toroidal configuration in stationary and nonstationary variants. It is now regarded as one of the most promising.

"The prospects are now better than ever before; several years ago Russian experiments invented an apparatus called a tokamak. . . . This system has been rather successfully reproduced in the U.S.A.," wrote Hans A. Bethe in 1976.[1]

"The cleverest and most promising device has been the so-called tokamak proposed in the USSR," said P. L. Kapitsa in his Nobel Prize lecture in 1978.[2]

A very complete picture of the present state of the problem of controlled thermonuclear synthesis reactions has been given by the associate director of the Division of Thermonuclear Research, De-

[1]Hans A. Bethe, "The Necessity of Nuclear Power," *Scientific American*, Jan. 1976.
[2]P. L. Kapitsa, "Plasma and the Controlled Thermonuclear Reaction" (Nobel Prize lecture), *Science*, Sept. 7, 1979.

partment of Energy of the U.S.A., John F. Clarke, in a survey written in December 1979.[3] Here are some excerpts from this article:

Recent experimental results from the United States, the USSR, Europe, and Japan indicate that the tokamak, one of a number of possible fusion approaches, can confine a plasma sufficiently well to produce power. . . there is no fundamental technological obstacle to translating the scientific success of tokamak development to the production of controlled fusion power. . . .
. . . We are encouraging joint planning of research on the world class tokamaks now under construction, the T-15 in the USSR, JT-60 in Japan, JE7 in Europe, and TFTR in the United States. . . . This program should prepare the ground for the next step—taking the fusion program into the engineering development phase—perhaps as early as 1981.

Sakharov has also been working on an alternative, radically different, method of magnetic isolation and containment of a plasma, involving the use of lasers. In his brief autobiography, A. D. Sakharov writes: "In 1961 I proposed for this same purpose [production of a controlled thermonuclear reaction] that deuterium be heated by means of a pulsed laser beam."[4] This idea was thought of independently in various countries and is now being intensively developed in the USSR and elsewhere.

Let us look in more detail at the physical nature and the significance of controlled thermonuclear synthesis.

The merging together or fusion of two deuterium nuclei or a deuterium nucleus and a helium nucleus produces nuclei of helium isotopes and fast neutrons. The positive energy yield is due to a decrease of the total rest mass of the reacting particles, in accordance with Einstein's famous equation $E = mc^2$. For nuclei to merge, they must approach each other to within the range of action of nuclear forces. This is hindered by the electrostatic repulsion, and to overcome this repulsion the nuclei must have a sufficiently large kinetic

[3]John F. Clarke, "The Next Step in Fusion," *Science*, Nov. 28, 1980.
[4]Andrei Dmitriyevich Sakharov, *Sakharov Speaks* (New York: Knopf, 1974).

energy of their thermal motion. Accordingly, to produce a thermonuclear reaction there must be a high initial ignition temperature.
In a hydrogen bomb the igniter is an atomic bomb. In the case of
controlled thermonuclear synthesis the necessary initial heating can
in principle be obtained by bringing about powerful electric dis-
charges in the deuterium or deuterium-tritium plasma. Here the main
problem is to maintain this "lightning" for the length of time (several seconds) necessary to heat the plasma to the ignition temperature of the thermonuclear reaction. It is also necessary that the energy released in the synthesis of nuclei be larger than that expended
in heating the plasma; only in this case can we say that "a fire has
been lit."

No walls made of matter can serve to contain the plasma, since at
such a high temperature they would be vaporized at once. The only
possible way to retain the hot plasma in a limited volume is the use
of very strong magnetic fields. A plasma is a gas of electrically charged
particles, whose paths of motion are curved by the action of a magnetic field. By a proper choice of the configuration of the external
magnetic field, with allowance for the "pinch" (the self-compression
of a stream of plasma particles by its own magnetic field), one can
hope to avoid the breaking out of the plasma through any wall. The
main difficulty that arises is the instability of plasma streams. There
are other difficulties also, for example breaking of the reactor's walls
by neutral atoms, which are always present in small numbers in a
plasma and of course are not restrained by the magnetic field. These
other problems that arise are, indeed, of technical rather than fundamental character.

In the very long time scale there are three energy sources that can
be used, namely solar energy, atomic (fission) energy, and thermonuclear energy. The disadvantage of the solar source is the low density
of the energy. The problems of atomic energy are the necessity of
burying radioactive wastes, and, greatest of all, the danger of the uncontrollable spread of atomic weapons in the world. Thermonuclear
synthesis is much safer as regards "slag" and also as regards "atomic
terrorists." Millions of tons of deuterium are contained in the world's
oceans. There is practically no free tritium in nature, but it can be produced in thermonuclear reactors themselves from lithium (helium and

tritium are the products of the interaction of neutrons with lithium nuclei). Accordingly, the only shortcoming of the controlled thermonuclear reaction is that up to now it has not been achieved.

In the scientific laboratories of many countries, broad investigations are now being conducted on various versions of the solution of the problem of a controlled thermonuclear reaction. Since Kurchatov's lecture at Harwell, which produced an enormous impression throughout the world, research on controlled thermonuclear reactions has been conducted openly and in close international cooperation. It has been an example of an entire system of international cooperation, founded in the 1950s to 1970s and subjected to the shock of well-known events, including the conviction of Yu. F. Orlov and the exile of A. D. Sakharov.

On September 14, 1981, the Tenth European Conference on Plasma Physics and Controlled Thermonuclear Synthesis will be opened in Moscow. Is such a conference possible without the participation of the one who laid the foundation of the whole process, Academician Sakharov? The illegal detention of Sakharov makes this an exceptionally sharp question. Moreover, it must be well known that the head of the Soviet organizing committee for the conference, Academician Velikov, has during the year just past repeatedly ignored Sakharov's appeals for aid.

2. Magnetoplosive Generators

In 1951–52 Sakharov suggested a principle for using the energy of an explosion to produce ultrastrong magnetic fields and ultrastrong currents. This principle is based on the conservation of magnetic flux and the increase of magnetic energy during rapid explosive deformation of metallic circuits carrying current, in particular during cumulative collapse of metal cylinders giving rise to the term "magnetic cumulation" (MC). This proposal of Sakharov's and the results of research done on his initiative were published in 1965 and 1966.[5] These

---

[5]A. D. Sakharov, R. Z. Lyudaev, E. N. Smirnov, Yu. N. Plyushchev, A. I. Pavlovskii, V. K. Chernishev, E. A. Feoktistova, E. I. Zharinov, and Yu. A. Zysin, "Magnetic Cumulation," Doklady Akad. Nauk SSSR (Reports of the Academy of Sciences of the USSR) 165, No. 1 (1965) [Transl.: Soviet Phys. Doklady 10, 1045 (1965)]; A. D. Sakharov, "Magnetoplosive Generators," Usp. Fiz. Nauk 88, No. 4 (1965) [Sov. Phys. Uspekhi 9, 294 (1966)].

papers report the production of a record magnetic field of 25 million gauss (i.e., an energy density a million times that in a good permanent magnet). Many foreign publications on this subject also appeared in the 1960s. Sakharov writes:

> The field of application of MC generators is the solution of problems in physics and engineering such as, for example, the development of comparatively small-scale single-stage accelerators for producing high energies (100–1000 BeV), the production and study of dense high-temperature plasmas, the acceleration of solid objects to speeds of hundreds and thousands of kilometers per second, which is necessary for the solution of some astrophysical problems (producing stellar temperatures and pressures under laboratory conditions), the physics of shock waves, the study of the equations of state and the properties of substances at ultrahigh temperatures and pressures, studying the effects of meteorites on the casings of spaceships, and so on.
>
> There are prospects of investigating the electric, optical, and elastic properties of various substances in such magnetic fields which hitherto were practically unattainable.

The papers by Sakharov and his collaborators describe two characteristic explosive generators: MK-1 (compression of an axial magnetic field) and MK-2 (expulsion of the magnetic field from a solenoid and its subsequent compression by coaxial walls).

From a theoretical point of view, the MK-1 system is the simpler of the two. A pulse of current in a solenoidal winding produces a magnetic field inside a hollow metal cylinder. Outside the cylinder there is a coaxial layer of an explosive substance. A converging shock wave is produced in this charge. The explosion is timed so that the compression of the cylinder has started at the instant of maximum current in the solenoidal winding, i.e., at the instant when the initial magnetic field is largest. The rate at which the wall of the cylinder is pushed in is more than 1 km/sec; the motion is stopped because of the counterpressure of the magnetic field. The strength of the field is inversely proportional to the area of the cylinder's cross section, since the magnetic flux in the cylinder remains constant. This is due to the phenomenon of electromagnetic induction; when the cylinder wall moves in the radial direction, induced currents flow in it, and according to Lenz's law they act so as not to let the field out of the inside region. With an initial field of 30,000 gauss the very first experiments achieved a field of a million gauss, corresponding to a decrease in the radius by about a factor six. Sakharov writes:

> The MK-2 generator is of especial interest for the production of strong fields and very large magnetic field energies (with conversion of about 20 percent of the energy of the explosive into field energy, with relatively large values of the magnetic field, up to 2 million gauss).
>
> The practical realization of an MK-2 system with good characteristics required prolonged research by a large team, mostly completed in 1956 (the first MK-2 was constructed in 1952, and in 1953 currents of 100 million amperes were obtained).

The use of MK-2 generators to provide very high initial fields in an MK-1 generator made it possible to produce an ultrahigh field of 25 million gauss.

A magnetic field exerts a pressure on any barrier beyond which it cannot penetrate. This magnetic field pressure can be used to compress a specimen from all sides and to study the properties of various substances at ultrahigh pressures under conditions of adiabatic compression, i.e., without shock heating to tens of thousands of degrees. Some of the most recent Soviet and American publications report on the production by magnetic cumulation methods of pressures of several million atmospheres with the specimen's temperature changing by not more than 300 degrees Celsius. Research has been done in this way on the properties of quartz and corundum, and also on hydrogen, in the attempt to obtain it in a metallic state.

Sakharov's Papers on Fundamental Problems of Physics

1. *Theory of Elementary Particles*

The theory of the elementary particles of which all matter is composed, and also questions of cosmology—problems of the origin and evolution of the world—have been considered in a number of papers by A. D. Sakharov. In this work Andrei Dmitriyevich advanced some remarkable ideas bearing on the most general physical principles. I shall try to give a general survey of these papers, without going deeply into technical details intelligible only to specialists.

Elementary-particle theory and cosmology are now in a period of tempestuous development. In I. E. Tamm's picturesque expression, they are at the forward edge of present-day physics. The gigantic progress of experimental techniques has made it possible to pile up a multitude of new experimental facts in a relatively short time, about twenty years. Some of them undoubtedly must be regarded as great discoveries, which force us to reexamine scientific ideas which had seemed to be very firmly established. We mention some examples. In cosmophysics: residual radiation, which is the remaining trace of processes that occurred in the first instants of the world's existence; the discovery of pulsars, neutron stars, and possibly the observation of black holes in double-star systems. In particle physics: nonconservation of CP parity, "charmed" particles, upsilon particles, heavy leptons. This list of examples is far from complete and is given only for illustration.

This flood of discoveries has greatly stimulated theoretical thought, forcing a fresh reconsideration of fundamental scientific problems. In this the development of theory has not been merely subordinate to experimental progress. If theory only followed after experiment, trying to explain new data, I think no scientific development would be possible. Theory develops according to its own inherent laws. Moreover, the theoretical picture of the world is often unconfirmed (and sometimes even contradicted) by the experimental data that exist at a given time. (There have been many examples of this. One of the most striking is the general theory of relativity.) Nevertheless a "good" theory survives and itself influences the direction taken by experimental work. In time, theory and experiment come more or less into agreement with each other.

As the result of this sort of development of science an extremely peculiar picture appears. On the one hand, very general principles are formulated to govern colossal multitudes of experimental phenomena. On the other hand, fundamental problems appear, whose solution decidedly alters the scientific picture. As a rule, such problems arise at points where fundamental principles come into contradiction with each other.

In particle physics the most general principles are conservation laws for various physical quantities. Sometimes the way to solve an extremely difficult problem is to renounce the exact and universal validity of a conservation law.

Modern cosmology is based on Einstein's general theory of relativity, which is a theory of the gravitational field. The structure of the gravitational field is expressed in terms of the geometry of the four-dimensional space-time, or, as one says, of the world. This gravitational approach reflects very deep properties of the gravitational field. It is well known that all bodies move along absolutely similar trajectories in the earth's gravitational field, independently of their masses. This shows that motion in a gravitational field is geometric in its nature. The theory of relativity is a far-reaching generalization of this fact. It follows from the theory that as long as a gravitational field is weak (for example, the earth's field is in this sense extremely weak), the geometry of the space differs little from Euclid's geometry. But when the field becomes strong, the geometry of the space changes radically. One of the conclusions from the theory of relativity is that the entire world as a whole originated in a "big bang" about 10 billion years ago. At the instant of the "big bang" and immediately afterward, the geometry of space was extremely different from what we see around us today. The physical conditions at that time were also extremely different from terrestrial conditions; the matter density and the temperature were enormous (theoretically infinite). Physical processes could occur of types that are absolutely impossible under terrestrial conditions, or even in the cores of the sun and of ordinary stars.

A very remarkable peculiarity of the present picture of the world is that the problem of the microworld (the theory of elementary particles) and the problem of cosmology (the theory of the world as a whole) intersect each other. The physical processes at the instant right after the "big bang" and now controlling the evolution of the universe are essentially determined by the laws established in the physics of elementary particles. At the nexus of these two poles of the science there appear exceptionally interesting scientific problems and results.

Some of these problems interested Andrei Dmitriyevich Sakharov. In solving them, he tried to base his work on the most general principles in the given branch of science. This is a very difficult method, but in return the results are incomparably more important. Sakharov's approach is a very physical one. He always tried to relate several various (and sometimes seemingly remote) physical aspects of the given set of phenomena to each other and to obtain the maximum number of results capable of being tested experimentally.

2. The Baryon Asymmetry of the Universe

Baryons form a large family of "elementary particles that possess a certain physical property, the presence of a baryon charge." The best-known examples of baryons are the proton and the neutron, of which all atomic nuclei are composed. The proton and neutron are assigned baryon charges of 1 each. When they interact, baryons can be converted into one another, but the possible interconversions are limited by the condition that the total baryon charge of the initial particles must equal the total baryon charge of the final product particles. This condition can be expressed as the law of conservation of baryon charge. So far no experiment has ever shown a violation of this law. For each baryon there is a corresponding antibaryon. Antibaryons also belong to the baryon family. The properties of an antibaryon are analogous to those of the baryon, but at the same time they are opposite in a certain sense. The baryon charge of an antibaryon is opposite in sign to that of the corresponding baryon.

It must be especially emphasized that there is no intrinsic property that enables one to distinguish a particle from its antiparticle. The particle-antiparticle relationship is symmetric. For example, proton and antiproton are each other's antiparticles. If all the particles in the world were replaced with their antiparticles, the resulting world would differ very little from the one we live in. In other words, the antiworld, composed of "antimatter," would be constructed in exactly the same way as our world composed of "matter."

However, if a particle interacts with its antiparticle (for example, in the reaction proton-antiproton, which has been studied thoroughly in the laboratory), the picture is entirely different. Since the baryon charges of these two particles are of opposite signs, so that the total baryon charge is zero, there is no rule forbidding the conversion of the baryon-antibaryon pair into light particles, electrons, neutrinos, and photons (light quanta). This process is called annihilation, resulting in the disappearance of the baryon-antibaryon pair. If there existed in our world bodies composed of antimatter, then when they came into contact with matter annihilation would occur with the release of a large amount of energy. There have been no astronomical indications of the existence of such phenomena. Thus we know with great accuracy from experience that there are no concentrations of antimatter in our universe. Of course, it could always be answered that antimatter is concentrated somewhere in remote corners of the universe and separated in space from matter. But even if this were so at the present time, still in the time immediately after the "big bang," when all the material of the universe was in a superconcentrated state, it is very hard to imagine such a separation of matter and antimatter. We are left with the proposition that there are no islands of antimatter in the universe, or in other words that the baryon charge of the universe is not equal to zero. This is the baryon asymmetry of the universe; despite the fact that the laws of physics allow the replacement of matter with antimatter, the universe consists of particles with baryon charges all of the same sign. The problem of the baryon asymmetry of the universe poses the question: How could this come about in the process of evolution? As we see, in this problem the fundamental laws of physics intersect. On the one hand, the law of conservation of baryon charge, and on the other, the ideas of the general theory of relativity, from which follows the model of an expanding universe which arose as the result of a "big bang." Of course, one possible answer is that "it was always that way," that the superdense stuff at the time of the "big bang" had exactly the same baryon charge that the universe has today. This, however, is an empty answer.

A much more interesting hypothesis is that the initial state of the universe had baryon charge zero, and that the presently existing baryon asymmetry came about as the result of certain physical processes in the course of the evolution of the universe. It is from this point of view that various aspects of the baryon asymmetry of the universe are considered in papers by Sakharov (see References 1,2,3,4). Andrei Dmitriyevich investigates two different hypotheses. The first (3) is that baryon charge is strictly conserved in nature, but that as a result of nonstationary processes in the superdense material produced in the "big bang," a separation of the baryon charges is possible, in which the positive charge gets concentrated in nucleons and the same amount of negative charge in certain hypothetical particles, neutral antiquarks (with baryon charge $1/3$). By hypothesis these antiquarks cannot be captured by nuclei. The entire surrounding space is filled with antiquarks. Their average density is three times that of nucleons (the total baryon charge of nucleons and antiquarks is equal to zero). It is easy to formulate properties for the antiquarks that do not lead to any contradiction with existing data. The paper discusses possible experiments to observe antiquarks. Such experiments are the most direct tests of the hypothesis.

The second hypothesis (1,4) is quite different from the first. It is assumed that the law of conservation of baryon charge holds only approximately. A specific model of an interaction which does not conserve baryon charge is introduced. This interaction leads to the decay of protons into light particles (in a particular case described, into μ mesons). It is shown in the paper that in the nonstationary process of expansion of the superdense material the proposed mechanism can give the now observed value of the baryon asymmetry. At the present time the effects of the new interaction are extremely small. For example, although this interaction leads to decay of the proton, so that protons are unstable, still the lifetime of the proton is so large that observation of proton decay is far beyond the range of present possibilities. In a following paper (2), Sakharov develops his hypothesis and connects it with the effect of nonconservation of CP parity, an extremely important effect observed experimentally in the decay of long-lived κ mesons.

The problem of the baryon asymmetry of the universe is at present one of the central problems uniting two most important branches of physics—elementary-particle theory and cosmology.

3. *Cosmological Models*

The law of universal gravitation, which asserts that all of the bodies in the world attract each other, is one of the most universal laws of nature. The fact that on small scales of size the properties of systems are not determined by gravitational interactions is due to the fact that this interaction is (in a certain sense) an extremely weak one. For example, the electromagnetic interaction between the proton and the electron in a hydrogen atom is many orders of magnitude stronger than their gravitational interaction. However, there is no interaction in the world except the gravitational one that is always attractive. Therefore, when we proceed from the consideration of small-scale phenomena to ever larger ranges of size and mass, the relative importance of gravitation increases. If we then regard the entire world as a physical system, then for this sort of system gravitation is utterly dominant, and all other interactions are relatively insignificant. It can be said that gravitation completely determines the structure of the world as a whole.

Einstein's general theory of relativity expresses the gravitational field in terms of the geometry of the four-dimensional world. This geometry is such that for small space-time regions it can differ extremely little from Euclid's geometry. Considered as a whole, however, the entire space differs radically from Euclidean space. Accordingly, a cosmological model describing the whole universe reduces essentially to the consideration of a non-Euclidean space having definite properties. It is through the geometrical characteristics of this space that the physical properties of the world as a whole are expressed.

The model now most widely used in research on cosmology is Friedman's model of the expanding universe. In this model there exists a singular point, the "big bang." This point corresponds to the instant of time $t = 0$. For values $t < 0$, space does not exist (in this model). In (5) Sakharov proposes the concept of cosmological models for which also there exists a singular point at $t = 0$, analogous to the "big bang," but unlike Friedman's model these allow the definition of physical quantities also for $t < 0$. Sakharov called such models cosmological models with reversal of time's arrow.

The basic idea of models with reversal of time's arrow is related to the solution of the "global paradox of reversibility" which was formulated in the last century. The point of this paradox is that all dynamical laws of physics are invariant under a change of the direction of time (the substitution $t \rightarrow -t$), while the equations of statistical physics do not have this invariance. The irreversibility of the laws of statistical physics is the essential basis of the second law of thermodynamics, which asserts that entropy increases with time. In cosmological models with reversal of time's arrow, one can eliminate the "global paradox of reversibility" and formulate the laws of dynamics and of statistical physics in such a way that they are invariant under change of the direction of time. (We note that this was quite impossible in the Friedman model, since in the framework of that model it only makes sense to speak of time values $t > 0$.) Thus the law of reversibility takes on the quality of a fundamental law of nature. Actually, this law has to be made a bit more complicated: Along with the change of the direction of time (the T transformation) one has to make a mirror reversal of space (P transformation) and a replacement of every particle with its antiparticle (C transformation). The result is that the law of TPC invariance is formulated as a fundamental law of nature.

It follows from TPC invariance that at the instant of the "big bang" (t = 0) the world was neutral with respect to all conserved charges. This puts in a very sharp form the problem of the baryon asymmetry of the universe.

An extraordinarily interesting idea about the topological nature of charges was stated by Sakharov in (6). It is that matter consists of "elementary charges" which are rather tricky topological space-time structures. Moreover, the topologies of the world for t > 0 and t < 0 are related by TPC invariance. These new and very radical ideas are unfortunately still inadequately developed. If they can be successfully developed, Einstein's dream of the reduction of physics to geometry (a truly rather tricky notion) will have been to some extent realized.

4. The Concept of the Zero Lagrangian

In three papers (7,8,9) Sakharov develops the idea of the zero Lagrangian, according to which the action function of physical fields appears as the result of interaction of these fields with the physical vacuum. Here the vacuum is regarded not as "empty space," but as a universal physical system. In the absence of external fields the vacuum is in its ground state. An external field produces a polarization of the vacuum, which results in a change of the action function of the vacuum. This change can be regarded as the action function of the given physical field. Starting from some general principles, one can develop a method for calculating the effective action for various physical fields, such as the gravitational, electromagnetic, and electron-positron fields. Besides giving a general approach in principle, these papers are interesting for their development of a new method for calculating quantum effects.

One paper (9) considers the question of generalizing the Einstein theory of gravitation by introducing a scalar field into the theory. (Various versions of such a scalar-tensor theory of gravitation have been discussed repeatedly in the literature.) Generally speaking, the introduction of a new field into the theory violates a general principle on which Einstein's theory is based—that of the equivalence of inertial and gravitational masses. Sakharov shows that if a scalar-tensor theory of gravitation is deduced from the zero-Lagrangian principle, one gets a special variant of the theory in which the equivalence principle is not violated, and along with this the scalar field becomes in principle unobservable. The theory so obtained is physically equivalent to Einstein's theory.

5. The Problem of the Masses of Elementary Particles

The mass of an elementary particle is one of a few fundamental properties of each particle, and therefore the finding of laws describing the masses of particles is one of the basic theoretical problems. It is true that the very concept of elementary particles from which all matter is constructed has turned out to be very deceptive. As soon as some type of particle, which might be regarded as elementary, has begun to be intently studied, a tendency has quickly appeared for the number of such "elementary" particles to increase; scientists have discovered many new "elementary" particles. That is the way it was with atoms. Later on, after the discovery of the proton and the neutron, these particles were accepted as elementary, and a picture of atomic nuclei made up of various numbers of protons and neutrons was constructed. However, as experimental technique was developed, many representatives of the baryon family of particles (to which proton and neutron belong) were discovered, and also a set of particles with baryon number zero—the mesons. By now some hundreds of these particles are known, and while they are still called "elementary," everyone understands that this is just a convention. In order to bring order into the "household" of elementary particles, various hypotheses have been proposed. One of the most successful hypotheses (still current today) was that all of these particles are composed of still more elementary particles called quarks. The original suggestion was that there are three types of quark (quite in agreement with James Joyce). It was assumed that all baryons are composed of three quarks, and mesons of a quark and an antiquark. By using various combinations of the three types of quarks it was possible to include in the system all of the "elementary" particles known at that time. It is true that rather soon three types of quark turned out to be too few for the explanation of the properties of all the new particles—one had to assume the existence of first a fourth and then a fifth quark. To this set (on purely theoretical considerations) a sixth quark was added. Physicists now hope that this set will suffice.

In a cycle of papers (10, 11, 12, 13) Sakharov used the quark hypothesis to construct formulas describing the masses of observed particles. These formulas do not follow from basic scientific principles, but are rather of a semiempirical nature. It is remarkable, however, that from extremely simple physical arguments, using a minimum of information about quarks and introducing a minimal number of unknown (adjustable) parameters, one can succeed in finding mass formulas that give good agreement with a very large number of experimental values.

Sakharov refines and perfects the mass formulas, bringing them into agreement with the rapidly increasing experimental knowledge in particle physics. When new "charmed" particles were discovered, requiring for their explanation the assumption that a fourth, "charmed" quark exists, Sakharov took this quark into account in his mass formulas (11). He succeeded not only in explaining well the masses of the newly discovered particles, but also in predicting to good accuracy the masses of particles not yet discovered, which soon were found experimentally. Subsequent improvements of the formula (12, 13) were stimulated by the progress of the theory (quantum chromodynamics). Sakharov gave his formulas a simpler form, reducing the number of parameters. The agreement with experiment was quite good.

6.

I believe that Andrei Dmitriyevich Sakharov's papers contain ideas of remarkable depth and originality. Only a few of them have received some development (this applies mainly to the problem of the baryon asymmetry).

I am sure that the development of these ideas will be extremely important for science. I hope it will occur in the near future.

References: Papers on Fundamental Problems by A. D. Sakharov

1. Violation of CP Invariance, C Asymmetry, and the Baryon Asymmetry of the Universe, *Pis'ma ZhETF* 5, 32 (1967) [English transl.: *JETP Lett.* 5, 24 (1967)].
2. Quark-Muon Currents and the Violation of CP Invariance, *Pis'ma ZhETF* 5, 36 (1967) [*JETP Lett.* 5, 27 (1967)].

3. "Antiquarks in the Universe." In collection dedicated to N. N. Bogolyu-
   bov on his sixtieth birthday. *Nauka*, 1969, p. 35.

4. "The Baryon Asymmetry of the Universe," *ZhETF 76*, 1172 (1979) [*Sov.
   Phys. JETP 49*, 594 (1979)].

5. "Cosmological Models of the Universe with Reversal of Time's Arrow,"
   ZhETF *79*, 698 (1980) [*Sov. Phys. JETP 52*, 349 (1980)].

6. "The Topological Structure of Elementary Charges and CPT Symmetry."
   In collection *Problems of Theoretical Physics* (collection in memory
   of I. E. Tamm), *Nauka*, 1972, p. 242.

7. "Vacuum Quantum Fluctuations in Curved Space and the Theory of
   Gravitation," Dokl. akad. Nauk SSSR *177*, 70 (1967) [*Sov. Phys.
   Doklady 12*, 1040 (1968)].

8. "The Spectral Density of Eigenvalues of the Wave Equation and Vacuum
   Polarization," *Teor. Mat. Fiz. 23*, 178 (1975) [*Theor. Math. Phys.*
   (USSR) *23*, 435 (1975)].

9. "On the Scalar-Tensor Theory of Gravitation," *Pis'ma ZhETF 20*, 189
   (1974) [*JETP Lett. 20*, 81 (1974)].

10. "Quark Structure and Masses of Strongly Interacting Particles," *Yad. Fiz.
    4*, 395 (1966) [*Sov. J. Nucl. Phys. 4*, 283 (1966)].

11. "Mass Formula for Mesons and Baryons with Effects of Charm In-
    cluded," *Pis'ma ZhETF 21*, 554 (1975) [*JETP Lett. 21*, 258 (1975)].

12. "Mass Formula for Mesons and Baryons," *ZhETF 78*, 2112 (1980) [*Sov.
    Phys. JETP 51*, 1059 (1980)].

13. "Estimate of the Constant Interaction of Quarks with the Gluon Field,"
    ZhETF 79, 350 (1980) [*Sov. Phys. JETP 52*, 175 (1980)].

Sakharov's
Problem Games

At various times A. D. Sakharov solved certain special physical and mathematical problems. We give a brief description of them, based on his own summary of his scientific work.

1. A cloud of rarefied gas with given equation of state is in the radiation field, in temperature equilibrium with the radiation, whose mean free path is much larger than the size of the cloud; the temperature of the radiation is a function of the time. A self-similar solution for the dispersal of the cloud is found.

2. A stream of viscous liquid emerges from a circular orifice and lengthens under the action of gravity. The shape of the stream is found at a distance from the orifice much larger than its radius. (Surface tension and inertia are neglected.)

3. On the plane interface between two transparent media there is a pigment that absorbs light. At the initial moment a beam of light of circular shape falls on the interface. The law of the temperature's increase at the center of the circle is found.

4. The force of electrostatic attraction between two convex conducting bodies whose minimum distance of separation is much smaller than their radii of curvature, for example two cylinders with their axes placed at an angle with each other, is found. The potential difference between the bodies is given. The problem arose in connection with an analogous problem in the theory of magnetism. While

working in a factory in wartime, Sakharov suggested a simple method for determining the thicknesses of nonmagnetic coatings on bullets, with a geometry similar to that for which the electrostatic problem is solved.

5. When cabbage is chopped with a cleaver, polyhedra with various numbers of vertices and of various sizes and shapes are produced. The average number of vertices and the ratio of the square of the mean perimeter of a polygon to its mean area are found. (Sakharov remarks: "The problem came up as a result of my chopping cabbage to help my wife make cabbage pie.")

6. A finite set of points on a plane is given. Each point is to be connected with a straight line with each other point, with the use of a given number of colors. Two theorems have been formulated (and partially proved) on the possibility of finding ways of coloring the connecting lines so that among the points there is no subset of n points (n being a given number) in which all of its points are connected by lines of the same color.

7. A circular vessel standing on a table contains a liquid. Several spots are made on the liquid's surface with ink. The vessel is turned by hand through some angle. It is shown that after the motion stops the same configuration of the spots is restored, turned through that same angle.

8. Two families of theorems relating to the theory of numbers are formulated. In particular: (1) The sequence $(n! + 1)$ contains an infinite number of prime numbers. (2) The sequence $[(n^2)! + 1]$ contains a finite number of prime numbers.

9. A rapidly converging algorithm for calculating the square roots of all integers and rational numbers is constructed, and also an algorithm for calculating terms of a Fibonacci series.

10. Simple approximate constructions for trisecting angles are devised.

11. A simple example is proposed of a hydrodynamical motion which leads to the "magnetohydrodynamic dynamo" effect.

Sakharov's Textbooks and Popular Science Articles

1. Sakharov worked along with M. I. Bludov on preparing new editions of a textbook on physics for technical schools written by his father, D. I. Sakharov, with Bludov's assistance. The first revised posthumous edition appeared in 1963. A newly revised edition was to have been published in 1974, but the permit for it was revoked after a newspaper campaign against Sakharov in 1974.

2. He prepared a problem book in physics compiled by D. I. Sakharov (last edition in 1974) for republication.

3. He wrote an article for the collection *The Future of Science*, edited by V. A. Kirillin, 1967. The collection did not go on sale. The article contains a forecast of the development of science and engineering. Some ideas from the article appeared in the paper "Thoughts on Progress, Peaceful Coexistence, and Intellectual Freedom."

4. He wrote an article, "The Symmetry of the Universe," in the collection *The Future of Science*, Znanie, 1967.

5. He wrote an article, "Does an Elementary Length Exist?" in the journal *Fizika v shkole*, 1969. It was an attempt to explain some ideas in local and nonlocal field theory in a popular manner.

VALERY SOIFER

Andrei Sakharov and the Fate of Biological Science in the USSR

Today it is hard to remember when I first heard of A. D. Sakharov. As best I recall, I was told about him in the middle 1950s by a friend of mine. And already, in the first account I heard, genuine respect was juxtaposed with a kind of mysteriousness when it came to Sakharov's personality. My friend had attended an important and very impressive conference of physicists involving the leading scientists of the country, many academicians, corresponding members of the academy, and doctors of science. Suddenly a rather thin young man came into the hall. No sooner had he appeared in an aisle than everyone in the hall—all those gray-haired VIPs of the world of physics—stood up and, turning toward the aisle, applauded as the young man went to take his place on the speakers' platform.[1]

"That was Sakharov," my friend told me. He said it with a certain breathiness in his voice, although in those days, in my opinion, he had no feelings of piety toward authorities and was inclined more toward nihilism than toward exaggerating the merits of others. I somehow never got from him any intelligible explanation as to why, specifically, the young Sakharov was shown such respect by his colleagues. In those years we often heard the names of I. E. Tamm,

---

[1] This episode is apparently one of those myths which grew up around the young scientist. Sakharov has stated that he never sat on the speakers' platform.—Ed.

L. D. Landau, P. L. Kapitsa, G. S. Landsberg, M. A. Leontovich, and V. A. Fok. Manuals and books written by those people were on sale, and their names were constantly mentioned on the radio and in the newspapers. One could sometimes see in magazines photographs of the round-shouldered Landau or the thickset Tamm, with his head slightly tilted. But the press maintained complete silence about the youngest academician, Sakharov. Even the sphere of his scientific interests was hidden from eyes and ears.

Then came the time when I transferred from the Timiryazevo Agricultural Academy, where I had studied for three-and-a-half years, to the first-year class in the physics department at Moscow University. A chair of biophysics had been established there in 1957; a friend and I joyfully seized upon the suggestion that we requalify ourselves in a specialty in which no one previously had been specially trained. But even there, in the physics department, Sakharov's name was not mentioned, and his articles were not offered to the students.

It may be that Sakharov's name first achieved wide currency among biologists rather than physicists.

Those were the years (1957–64) of the bitter struggle against Lysenko and other representatives of Michurinian biology who had set our science back decades, destroying the once-renowned traditions of the Russian geneticists, who had made a great contribution to world science.

From the moment of Lysenko's accession as the sole and unchallengeable leader in Soviet biology, all serious experimental and theoretical work in the field of genetics had stopped. Crucial damage was done, and with the years the losses to agriculture increased. The gang of Lysenkoites bent every effort to seize all positions in science and associated fields. This cancerous tumor grew and metastasized sometimes in one place in the huge body of the biological discipline, sometimes in another.

Lysenko flatly denied the significance of genes as material storers of hereditary information, asserting that "every grain of living matter possesses heredity." Therefore he groundlessly affirmed that no specific mutagenes—that is, substances selectively affecting the hereditary structures—exist or can exist; that organisms change as a whole, in parallel with a change in the external environment, and any favorable changes in the body are inherited. It is quite natural that in denying the existence of hereditary structures and the factors acting on them, Lysenko should have obstructed the experimental study of those problems, replacing genetics with pure quackery.

Paradoxical as it may seem, the "healing" of Soviet biology began from the outside, from a direction not under the control of Lysenko. The rebirth of genetics, and later of many other biological disciplines, was assisted by physicists.

With the development of nuclear physics, with the beginning of tests of atomic and nuclear weapons, a frightful peculiarity became apparent: the damaging of hereditary structures (genes) owing to the irradiation of living organisms. The first atomic physicists, knowing nothing of this, fell victim to the demon awakened by their own minds. All those who came into contact with fissionable substances died, slowly but surely, from radiation sickness. The agonizing death of the first nuclear physicists[2] was the price paid for ignorance of the laws governing damage to hereditary structures by radiation. But then, once having grasped the first laws of the influence of irradiation on chromosomes, the geneticists, together with physicists, began intensified work on a thorough study of those laws. Radiation genetics—an offspring of both biologists and physicists—began to develop at a very rapid pace. Life itself demanded the intensified study of genetic laws.

At that point, Soviet physicists proved to be the force that helped to revive these investigations despite the taboo of the Lysenkoites. A laboratory of radiation genetics was set up at the Biophysics Institute of the USSR Academy of Sciences. A radiobiological section was organized at the Institute of Atomic Energy at the initiative of I. E. Tamm, supported by I. V. Kurchatov; and similar laboratories were set up at several other physics institutes. Of tremendous importance was the organization of scientific seminars at which biological problems were considered—especially the theoretical seminars of I. E. Tamm at the Lebedev Physics Institute of the USSR Academy of Sciences.

At this stage, Andrei Dmitriyevich joined in the general work of

[2]In the first decades of the twentieth century, many doctors—roentgenologists and radiologists—fell victim to high levels of radiation. Therefore, the physicists were better prepared to work with large doses of radiation.—Ed.

physicists for the rebirth of research on radiation genetics. Then he began—with conviction, and with specifically Sakharovian thoroughness—his struggle against Lysenko.

Much of the history of that struggle has now been lost. Many important landmarks on the path to the rebirth of genetics in the USSR have remained unnoticed, and a good deal has been done quite deliberately so that no traces of it should remain. There are no stenographic records of several important speeches, and the number of surviving participants in the struggle against a monopoly of biology in the USSR is steadily dwindling. This makes it all the more imperative that we begin at once, without further delay, to gather materials and reminiscences which can still restore the history of those relatively far-off days.

Nonetheless, there is much that has not been forgotten. In 1959, Atomizdat published a slim collection of articles called *Soviet Scientists on the Danger of Testing Nuclear Weapons,* with a foreword by I. V. Kurchatov. The main article was one by Sakharov titled "The Radioactive Carbon of Nuclear Explosions and Nonthreshold Biological Effects" (pp. 36–44). The problems posed in that article were of basic significance.

For the majority of scientists, the damage to genes from high-level irradiation (including that in areas of explosions of nuclear devices) was obvious. But the question of those supposedly negligible traces of fissionable substances that were carried through the earth's atmosphere and the hydrosphere, and that only slightly increased background radiation, was more than a purely scientific problem. Many nuclear physicists quite simply denied the harmful effects of such radiation. The well-known physicist Edward Teller—the father of the American hydrogen bomb—even cynically declared that the damage from testing was "the equivalent of smoking one cigarette twice a month."

Andrei Dmitriyevich decided to analyze the problem thoroughly. His research demanded a clear understanding of the various aspects of the action of radiation on living matter, both on the organism as a whole and on the structures of heredity. That analysis razed to the ground the frivolous analysis of Teller and his followers. The calculation alone of neutron radiation with the formation of the long-lived radioactive carbon isotope C^{14} provided clear proof of the damaging effect on hereditary structures of "residual radiation" and radiation at the moment of the explosion. Andrei Dmitriyevich was the first to make a strict mathematical calculation of the disturbance of the hereditary apparatus of cells from neutron action, and to consider the various consequences of irradiation. In a very concise way he showed the role of mutations in the appearance of hereditary disease, the possibility of an increase in cancerous diseases and leukemia from irradiation, the decrease in the immunological reactivity of organisms, the damage done to man because of an increase in the mutability of microbes and viruses, and the associated periodically arising pandemics (outbreaks) of new forms of pathogenic viruses and bacteria. It was precisely that breadth of the consideration of biological laws, in combination with a strict mathematical calculation of the doses and a physical analysis of the process of gene damage—allowing for the growth in the world's population—which constituted the unique and most important part of Sakharov's paper.

Not yet discovered at that time was the property of living cells of eliminating part of the damage, restoring insofar as possible the original gene structure—the capacity for "self-healing," or, as the biologists say, for repairing one's own genes. Therefore, Andrei Dmitriyevich could not introduce corrections for reparation into his calculations. But even today, when those coefficients are taken into account, his calculations remain fully valid.

Thus the value of this research by Sakharov consisted in the fact that he was the first rigorously to combine data from strictly physical radiation with various biological data. This synthesis led to an impeccably demonstrated conclusion as to the damage caused by testing a nuclear weapon—that weapon which, to a considerable extent, was his own offspring.

To raise his hand against his own invention, to carry on a struggle for its banning—that was humanism in action. That was the highly moral position of a truly honest scientist, universally regarded as the most illustrious among the Soviet scientific elite, thrice Hero of Soviet Labor.

Already at this time—in a clear challenge to those who, like Teller and Latter, considered that "mutations (hereditary diseases) should be welcomed as a necessary sacrifice to the biological progress of the human race"—Andrei Dmitriyevich, expressing concern for the future of the earth, declared (in contrast to several of his colleagues):

I am inclined to regard uncontrollable mutations as an evil, as an additional cause of the death of tens and hundreds of thousands of people as a result of experiments with nuclear weapons. . .

One of the arguments of those who claim that testing is "harmless" is that cosmic rays produce greater doses of irradiation than the doses from testing. But that argument does not cancel out the fact that the sufferings and deaths of hundreds of thousands of people—including those in neutral countries and in future generations—are being added to the sufferings and deaths that already occur in the world. Two world wars also added less than 10 percent to the mortality of the twentieth century; but that does not make war a normal phenomenon.

Another argument widely occurring in the literature of several countries is that the progress of civilization and the development of new technology leads in many cases to the sacrifice of human lives. The example of lives lost because of the operation of automobiles is often cited. But that analogy is neither accurate nor legitimate. Automotive transport improves people's living conditions, and it leads to misfortunes only in individual cases as the result of negligence by specific individuals, who bear criminal responsibility for it. But the misfortunes caused by tests are an inevitable consequence of each explosion. In my opinion, the only specific trait in the moral aspect of the given problem is the total impunity that attaches to the crime, because in each individual case of a person's death it is impossible to prove that the cause was radiation; and also because of the complete defenselessness of descendants with regard to our actions.

Having investigated in essence the effect of radiation on heredity, Andrei Dmitriyevich was able, in addition, to clarify for himself the damage being done by Lysenkoism, which was obstructing the study of genetic laws. And once having achieved this clarification, he boldly joined in the battle against Lysenkoism and the Lysenkoites. This was especially manifested during the attempted election of people close to Lysenko—N. I. Nuzhdin and G. V. Nikolsky[3]—to the USSR Academy of Sciences. Lysenko was bending every effort to secure their admission to the Academy, using all legitimate and illegitimate means. The votes were cast secretly, and in two stages. First the candidates were put up for election in the respective sections of the Academy: the physicists in the physics sections, the chemists in the chemical, and the biologists in the biological. In the biology section, Lysenko, using his "voting machine"—that is, the docile majority he had planted there over the course of decades—rather easily got what he wanted; his candidates won recommendation. Now it remained only for the general meeting of all academicians automatically to confirm (also by secret ballot) the candidates who had passed through the sieve. Usually, the stumbling block in the elections was the voting in the sections. Who better knew the true value of a scientist than his colleagues? But in this case things turned out differently. A measure of responsibility for the purity of the title "member of the USSR Academy of Sciences" fell on the shoulders of those who had a greater moral right to it.

Tamm and Sakharov joined in the general debate at the session of the Academy of Sciences, and were able to demonstrate, rigorously and by force of argument, the spiritual and scientific worthlessness of the Lysenkoites, and to convince other members of the Academy that neither of the candidates backed by Lysenko could meet the high criteria for the title of academician. The general meeting blackballed both Nuzhdin and Nikolsky.

By this time Tamm and Sakharov had become legendary figures in biological circles. They complemented each other very well. Tamm prepared carefully for his speeches, discussing with various biologists the key points in his characterization of the Lysenkoites. His passionate speech on the antiscientific attitude of the Lysenkoites was supplemented by Sakharov's remarks, made in a calm tone but full of humor.

---

[3]Sakharov spoke only once, and only against Nuzhdin.—Ed.

But at the next elections the pressures became more severe. Khrushchev, Lysenko's protector, sternly warned the then president of the Academy, M. V. Keldysh, that if Nuzhdin again failed to win election, he, Khrushchev, would order the Academy dissolved.

The question of Nuzhdin was taken up at a session of the Politburo, and he was given a special place. (It should not be forgotten that in the USSR, election to the Academy brings with it strong financial support in the form of a monthly honorarium, the right to free use of an automobile, sanatoria, etc.) Khrushchev's attitude was that if the Central Committee had approved a candidate, the Academy was obligated to elect that candidate.

Under such conditions one had to have a certain courage to speak out against the personal protégé of the head of the Party. And a very real threat hung over the whole Academy.

Despite everything, Sakharov stuck by his convictions. After several speeches by genuine scientists, including Sakharov, Nuzhdin was once again blackballed. The dissolution of the Academy of Sciences was prevented only by Khrushchev's removal from office just after that ill-starred session. Promptly thereafter, Lysenko himself suffered disaster.

It is hard to say what would have been the result of a policy of compromise in this matter, and for how many years the upswing of the biological sciences in the USSR would have been held back after the thirty-year dominance of the Lysenkoites.

It seems to me that this struggle by Andrei Dmitriyevich for the interests of science turned out to be basically important for him personally. That struggle brought to light something in him that singled him out from among many colleagues: a capacity for public activity, a lack of fear in the face of pressure of any kind, and adherence to principles in everything. In the years when he was speaking out in public as an opponent of Lysenkoism, he had not yet proved to be a fighter for the ideals of humanism—something that brought him world recognition. This was perhaps his first test of strength. But it was a test that clearly showed the strength of this amazing man.

In conclusion, I should like to tell about one of my meetings with Igor Evgenyevich Tamm, who gave me tremendous, invaluable help in my life in that he recruited me for work in the field of biophysics in 1957. From that time on, I often met with him at his home to discuss new developments in biology. One evening just before bedtime when we went for a walk along the embankment in front of the building in which he lived, he suddenly told me about one of his mistakes.

Tamm had a unique way of talking. The words came out of him rapidly, like bursts from a machine gun. It was the vivid, picturesque speech of one of the outstanding thinkers of our time, and his stories were unforgettable. Here is how I remembered that particular story:

"You know," he began, "it's not often that I make mistakes about people, especially beginning students. When I was teaching, quite a few talented youngsters passed through my hands, and they could be sized up rather quickly. But once Dau[4] brought me a student, Sakharov—he was in his third year, as I recall—and suggested that I work with him individually.[5] We sat down to talk, and talked for a long time. 'You know, young man,' I said to Sakharov, 'it's hardly likely that you'll make a real physicist. You have a sort of humanistic cast of mind.' But three years later, Sakharov and I made the hydrogen bomb. That," concluded Tamm, "was the mistake I made."

Tamm was convinced that he had made a mistake then. But as we can now see, even in this respect the great Tamm was equal to the occasion. Undoubtedly, Sakharov became an outstanding physicist. But time has also confirmed how right Tamm was in his first evaluation of Andrei Dmitriyevich's inner nature. Sakharov's talent has proved to be many-sided. He became an outstanding theoretical physicist and, at the same time, the greatest humanist of our age.

[4]An affectionate nickname for Academician Lev Davidovich Landau, used by his friends and colleagues.
[5]In fact, Sakharov first met Tamm in 1945, after graduation from the university and work at a plant. He met Landau much later.—Ed.

HERBERT F. YORK

Sakharov and the Nuclear Test Ban

Since 1958, the stated policy of both the United States and the Soviet Union has been to achieve a treaty banning all nuclear weapons tests. Andrei Sakharov, we now know, played a major, perhaps crucial, role in this matter. However, during the formative period—that is, the late 1950s and early 1960s—those of us involved in making and carrying out U.S. policy in this matter had no idea who Sakharov was or what he was doing.

I believe we would all have done a much better job than we actually did if we had at that time known more about him, his ideas, and his activities. But the curtain of secrecy with which the Soviets habitually surround most matters prevented that.

The policies of both countries regarding a test ban grew out of two quite distinct origins. One was the belief, held mainly by nuclear specialists, that the nuclear arms race was in itself a grave threat to mankind. The other was a narrower (and shallower) but much more widely shared concern about the dangers of radioactive fallout produced by nuclear weapons tests.

Concern about the nuclear arms race itself dated from the last years of World War II, even before the atomic bombs were dropped. The Bohr memorandum, the Szilard petition, the declaration by Harry Truman, Clement Attlee, and MacKenzie King in Washington in December 1945, the proposals by Oppenheimer, Lilienthal, and others that eventually became the Baruch plan—all are manifestations of this profound concern.

Early proposals for solutions to the problem were generally very broad in nature: some called for internationalizing the atom and total elimination of nuclear weapons; others, even broader, called for general and complete disarmament. The extreme nature of these proposals, as well as the generally worsening "Cold War" political climate in the late 1940s and early 1950s, rendered them unachievable.

Concern about the fallout problem dates very largely from the "Bravo" nuclear test of March 1, 1954. The Bravo test involved the largest nuclear explosion to that date, 15 megatons. Because of a slight shift in the wind, it resulted in a fallout pattern which placed lethal and near-lethal levels of fallout on and near a Japanese fishing vessel, the *Fortunate Dragon,* and on a number of inhabited islands. Only one person died as a result, but a further slight shift in this wind pattern would have killed hundreds.

There was a worldwide anti-test and anti-war reaction. Linus Pauling was one of its intellectual leaders, as was Albert Schweitzer. And others included the small group that signed the Russell-Einstein Manifesto calling for the abolition of war, and implicitly referring to this specific fallout incident as its immediate stimulus. The fallout issue even directly entered into the 1956 U.S. presidential campaign, when the Democratic candidate, Adlai Stevenson, drew attention to the problem and endorsed the idea of eliminating nuclear tests.[1]

The interplay of these two distinct sets of ideas—concern over the nuclear arms race and concern over fallout from nuclear tests—was nurtured by the general (though gradual) improvement in climate that followed the death of Stalin. Eventually, the idea of a test ban as a separable and feasible "first step" took hold, and serious studies of various aspects of that particular approach were initiated in both East and West.

Premier Nikita Khrushchev in 1957 issued several calls for a test

[1]Good accounts of the reaction to Bravo are provided in Robert Gilpin, *American Scientists and Nuclear Weapons Policy* (Princeton, N. J.: Princeton University Press, 1962), pp. 137–43, 154–61, 164–71; Lawrence S. Wittner, *Rebels Against War: The American Peace Movement, 1941–1960* (New York: Columbia University Press, 1969), pp. 235–37, 240–47.

ban, and in 1958 arranged for the Supreme Soviet to issue a decree banning all nuclear tests in the Soviet Union.[2] President Eisenhower in 1958 turned for help on this issue to his Science Advisory Committee, just after it was enlarged and its status elevated as a response to Sputnik. He asked two questions: Would a nuclear test ban be in the best interests of the United States? Could a test ban be adequately monitored? The conclusions of the committee, which held a special meeting in Puerto Rico, remote from telephone calls and other interference, was "yes" to the first question and a tentative "yes" to the second. (Among the seventeen members of the committee who were especially interested in this problem then and who stayed with it for years afterward were the chairman, James Killian, and George Kistiakowsky, Jerome Wiesner, Hans Bethe, I. I. Rabi, and myself.)

A bilateral Conference of Experts, held in Geneva in the summer of 1958, was chaired on our side by James Fisk, the president of Bell Laboratories. It produced the outlines of a verification system which came to be known as the Geneva System. In the fall of that same year, a Political Conference was opened, with the assigned task of producing a formal treaty, and a moratorium on all nuclear weapons tests was instituted at the same time. The purpose of this moratorium was to provide the proper climate for working out a treaty.

On the U.S. side, many of those who contributed to formulating nuclear arms control policy, who were on the negotiating team, or who were backing it up at home were directly involved in nuclear weapons development: Hans Bethe, Harold Brown, Ernest Lawrence, and myself, among others. None of the Soviets we dealt with directly seemed, at the time, to have such close connections to their weapons program. We now know that major participants in the Soviet program were very much involved. In Sakharov's words:

> Beginning in 1957 (not without the influence of statements on this subject made throughout the world by such people as Albert Schweitzer, Linus Pauling, and others) I felt myself responsible for the problem of radioactive contamination from

[2]U.S. Department of State, *Documents on Disarmament, 1945–1959* (Washington, D.C., 1960), II, 978–80.

nuclear explosions. As is known, the absorption of the radioactive products of nuclear explosions by the billions of people
inhabiting the earth leads to an increase in the incidence of
several diseases and birth defects, of so-called sub-threshold
biological effects—for example, because of damage to DNA
molecules, the bearers of heredity. When the radioactive
products of an explosion get into the atmosphere, each mega-
ton of the nuclear explosions means thousands of unknown victims. And each series of tests of a nuclear weapon (whether
they be conducted by the United States, the USSR, Great
Britain, China, or France) involves tens of megatons, i.e., tens
of thousands of victims.

In my attempts to explain this problem, I encountered great
difficulties—and a reluctance to understand. I wrote memo-
randums (as a result of one of them I. V. Kurchatov made a
trip to Yalta to meet with Khrushchev in an unsuccessful attempt to stop the 1958 tests), and I spoke at conferences.[3]

Sakharov Speaks, the book from which that quotation comes, was
published well after the events and then only in the West, but it
turns out that there was at least one contemporary source, published
in Moscow, which basically confirms what Sakharov said. That book,
Soviet Scientists on the Danger of Nuclear Tests, was published in
Moscow in 1960, in both Russian and English. (I first saw the Russian
version in Stockholm in 1970, and was not aware of the English translation until 1981!) In it, there is a statement by Kurchatov, the scientific head of the Soviet nuclear weapons program since the early
1940s, and an essay by Sakharov, both of which touch on the issue.
Kurchatov wrote:

When the war was coming to a close . . . United States
aircraft dropped two atomic bombs on the Japanese towns of
Hiroshima and Nagasaki, killing over 300,000 persons. . . .
The United States military politicians took advantage of these

[3]Andrei D. Sakharov, *Sakharov Speaks,* ed., Harrison E. Salisbury (New York: Alfred
Knopf, 1974), p. 32.

bombings to launch on a course of power politics against the USSR.

Soviet scientists regarded it as their sacred duty to ensure the safety of their country, and, under the day-to-day guidance of the Party and Government, together with the entire nation, they achieved outstanding success in the building of atomic and hydrogen weapons. Now, he who dares take the atomic sword against the Soviet people shall perish with this sword.

But the very thought of nuclear warfare is horrifying. We scientists working in the field of atomic energy see more clearly than anyone else that a war with atomic and hydrogen weapons would inflict incalculable suffering on humanity.

. . . Our scientific community has unequivocally called for a ban on nuclear weapons. This is also the stand taken by such world-famous scientists as Niels Bohr (Denmark), Linus Pauling (the United States of America), Heisenberg (Germany), Yukawa (Japan), Powell (Great Britain), the late Joliot-Curie (France), and many others.

Tests of atomic and hydrogen weapons not only hold the world in the grip of constant anxiety as the portent of a possible future atomic war, but are (and in future will be still more) a hazard to the health of humanity. . . .

We sometimes hear it said that it is not always possible to determine whether a given country is conducting nuclear bomb tests. This is hollow and irresponsible talk! It is common knowledge that there are many methods for detecting explosions of atomic and hydrogen bombs at very great distances. Among them are, for instance, the study of seismic oscillations, infrasonic waves, and the radioactivity of the atmosphere. The Geneva conference of experts (August 1958) demonstrated that such control is in every way feasible and realistic.[4]

The basic appeal for restraint and the "required" anti-American

[4]A. V. Lebedinsky, ed., *Soviet Scientists on the Danger of Nuclear Tests* (Moscow: Foreign Languages Publishing House, 1960), pp. 5–7.

statement are both noteworthy. Most of the other essays contain similar statements. But the Sakharov essay stands out from the rest in totally ignoring this standard bow towards the official polemics. In his contribution, Sakharov first estimates the amount of radioactivity that a large thermonuclear explosion would produce and then goes on to estimate the genetic and other biological damage such radioactivity would ultimately cause. He then asks, "What about the political and moral issue?" He answers his own question:

> One of the arguments of those who support the theory of the harmlessness of nuclear tests is that cosmic rays produce larger doses of radiation than do the tests. But this argument does not eliminate the fact that added to the existing distress and death of human beings are the death and distress of hundreds of thousands more, and these include people in neutral countries and in future generations. Two world wars also added less than 10 percent to the death rate in the twentieth century, but this fact does not make wars a normal occurrence.
>
> Another widespread argument in the literature of a number of countries is that the progress of civilization and the development of technology lead to loss of human life. A common example used is automobile accidents. But the analogy here is neither exact nor justified. Motor transport improves the living conditions of the population, while accidents are the result of carelessness on the part of specific individuals who are held responsible. The distress due to testing, on the contrary, is an inevitable consequence of each explosion. It is the author's opinion that, ethically speaking, the only peculiarity of this problem is the total impunity of the crime (for in no concrete case can it be proven that the death of a person is caused by radiation) and also the total defenselessness of future generations with respect to our actions.
>
> The cessation of test explosions will preserve the lives of hundreds of thousands of people and will have a still greater indirect effect by helping to lessen international tension and to

reduce the possibility of a nuclear war—the greatest danger of our age.[5]

The nuclear test moratorium of 1958–1961 was denounced after being in place only fourteen months (that is, at the very end of 1959), first by Eisenhower and then immediately afterward by Khrushchev. Eisenhower's reason was that it had originally been intended to run for only one year, during which time a formal treaty was supposed to be completed; but that had not happened. Eisenhower did not, however, authorize the resumption of nuclear tests. He simply declared that the United States was no longer bound by the moratorium, but would not resume tests without giving prior notice.[6] In his counter-denunciation, Khrushchev added a more restrictive codicil, namely, that the USSR would not resume unless the West did so first.

Despite these paired denunciations, the remarkable fact is that the moratorium continued in effect for another year and a half. As far as the United States was concerned, the main reason for the continuation of the moratorium was Eisenhower's deeply felt view that it was essential to find some way to contain the nuclear arms race, and that a nuclear test ban could contribute to that goal. He was greatly supported in this very controversial policy by George Kistiakowsky, then his science advisor, as well as by James Killian, Jerome Wiesner, myself (I was then head of research and engineering in the Pentagon), and others.

The moratorium also continued in effect for some time in the Soviet Union, again despite the double denunciation. In the summer of 1961, however, the Soviets took the initiative in ending it after almost three years with a long series of nuclear tests, followed some months later by an American response in kind.[7]

What was going on in the Soviet Union during the moratorium? *Sakharov Speaks* provides one answer:

> I remember that in the summer of 1961 there was a meet-
> ing between atomic scientists and the chairman of the Council

of Ministers, Khrushchev. It turned out that we were to prepare for a series of tests that would bolster up the new policy
of the USSR on the German question (the Berlin Wall). I wrote
a note to Khrushchev, saying: "To resume tests after a three-
year moratorium would undermine the talks on banning tests
and on disarmament, and would lead to a new round in the
armaments race—especially in the sphere of intercontinental
missiles and antimissile defense." I passed it up the line.
Khrushchev put the note in his breast pocket and invited all
present to dine. At the dinner table he made an off-the-cuff
speech that I remember for its frankness, and that did not
reflect merely his personal position. He said more or less the
following: Sakharov is a good scientist. But leave it to us, who
are specialists in this tricky business, to make foreign policy.
Only force—only the disorientation of the enemy. We can't
say aloud that we are carrying out our policy from a position
of strength, but that's the way it must be. I would be a slob,
and not chairman of the Council of Ministers, if I listened to
the likes of Sakharov.[8]

For a confirmation of Sakharov's words, but from a very different
perspective, let me turn to *Khrushchev Remembers*.

Literally a day or two before the resumption of our testing
program, I got a telephone call from Academician Sakharov.
He addressed me in my capacity as the Chairman of the Council of Ministers, and he said he had a petition to present. The
petition called on our government to cancel the scheduled nuclear explosion and not to engage in any further testing, at
least not of the hydrogen bomb. "As a scientist and as the
designer of the hydrogen bomb, I know what harm these explosions can bring down on the head of mankind."

I replied, "Comrade Sakharov, believe me, I deeply sympathize with your point of view. But as the man responsible
for the security of our country, I have no right to do what

[8]*Sakharov Speaks,* p. 33.

you're asking. For me to cancel the tests would be a crime against our state. I'm sure you know what kind of suffering was inflicted on our people during World War II. We can't risk the lives of our people again by giving our adversary a free hand to develop new means of destruction. Can't you understand that? To agree to what you are suggesting would spell doom for our country. Please understand that I simply cannot accept your plea; we must continue our tests."

My arguments didn't change his mind, and his didn't change mine; but that was to be expected. Looking back on the affair, I feel Sakharov had the wrong attitude. Obviously, he was of two minds. On the one hand, he had wanted to help his country defend itself against imperialist aggression. On the other hand, once he'd made it possible for us to develop the bomb, he was afraid of seeing it put to use. I think perhaps he was afraid of having his name associated with the possible implementation of the bomb. In other words, the scientist in him saw his patriotic duty and performed it well, while the pacifist in him made him hesitate. I have nothing against pacifists—or at least I won't have anything against them if and when we create conditions which make war impossible. But as long as we live in a world in which we have to keep both eyes open lest the imperialists gobble us up, then pacifism is a dangerous sentiment.

This conflict between Sakharov and me left a lasting imprint on us both. I took it as evidence that he didn't fully understand what was in the best interests of the state, and therefore from that moment on I was somewhat on my guard with him.[9]

After the resumption of testing, the world situation took several surprising turns, the most critical of which was the Cuban missile crisis. The nuclear scare that grew out of this confrontation stimulated a renewed effort to contain the nuclear arms race, and the Limited Nuclear Test Ban of 1963 resulted the very next year. This treaty in

[9]Nikita Khrushchev, *Khrushchev Remembers: The Last Testament*, translated and edited by Stobe Talbot (Boston: Little, Brown, 1974), pp. 69–70.

effect finessed what had been one of the main blocks to accomplishing a treaty earlier. That block was the especially difficult problem of monitoring a ban on underground tests. We now know that Sakharov personally played a role in breaking this particular bottleneck:

> Talks on the banning of nuclear testing had already been going on for several years, the stumbling block being the difficulty of monitoring underground explosions. But radioactive contamination is caused only by explosions in the atmosphere, in space, and in the ocean. Therefore, limiting the agreement to banning tests in these three environments would solve both problems (contamination and monitoring). It should be noted that a similar proposal had previously been made by President Eisenhower, but at the time it had not accorded with the thinking of the Soviet side. In 1963 the so-called Moscow Treaty, in which this idea was realized, was concluded on the initiative of Khrushchev and Kennedy. It is possible that my initiative was of help in this historic act.[10]

The moratorium of 1958–1961 and the Limited Test Ban of 1963 did lead to a large reduction in the radioactive pollution of the atmosphere (French and Chinese tests continued, but at a much lower level). They were also successful as a "first step" in the total attempt to contain the nuclear arms race. Other arms control treaties did indeed follow them: the Ban on Weapons of Mass Destruction in Outer Space, the Non-Proliferation Treaty, SALT I and SALT II and others were in fact built on the precedents created in working out the moratorium and the Limited Test Ban.

In the years since the events I have described, Sakharov has gone public with his effort to slow and reverse the nuclear arms race. In his later writings, he continued to stress the absolute need for doing something to curb the nuclear arms race, no matter how difficult it might be. . . .

Sakharov has also pointed out what is admittedly one of the most serious problems confronting those who advocate nuclear arms con-

[10]*Sakharov Speaks*, p. 34.

trol measures. That is the Soviet (and Russian) penchant for secrecy, and the concomitant repressive actions which keep the Soviet people from knowing much about what their own government is doing, and from acting on what little they may know. . . .

This is indeed one of the very important difficulties confronting us, but not the only one. There are plenty of regressive elements in our own domestic situation which also have a powerful negative impact on our attempts to contain the nuclear arms race. In fact, the blame for the failure to negotiate a test ban during the Carter administration lies somewhat more heavily on Washington than on Moscow. Ultimately, of course, the Soviet invasion of Afghanistan made completing such a treaty impossible, but the sorry fact is that the Carter series of comprehensive test ban negotiations were essentially moribund for a year before that happened.

If we look at the restraints on the nuclear arms race that have actually been achieved, as compared with the difficulty of getting anything done at all, we can say that much has been accomplished. But if the comparison is between what has been done and what needs to be done in view of the danger, we must admit that our achievements are pitifully inadequate. During the Carter administration, the effort to achieve a ban on all nuclear tests was given new life, and other, similar initiatives were undertaken. But nothing resulted.

The forces of reaction in both capitals proved to be too powerful. They fed on each other, with the result that the forward momentum of the process had been largely lost even before Tehran, Afghanistan, and the latest U.S. elections. But despite the setbacks and the sorry record, we must resolve to continue the work to contain the nuclear arms race—an effort started by Niels Bohr, Leo Szilard, Robert Oppenheimer, Linus Pauling, and Eisenhower and his committee in the West; by Peter Kapitsa, Andrei Sakharov, and other still unknown persons in the East.

We must follow the example of these noble men by dissipating the smoke screen that others place around these vital issues in order to confuse us. We too must learn how to discover which are the really important problems that confront us; and above all, we must have the courage to go where our convictions send us.

PHILIP HANDLER

Human Rights and Scientific Interchange

On the occasion of the International Conference in Honor of
Andrei Sakharov, Caspary Auditorium, Rockefeller University,
New York City, May 2, 1981.

The occasion of Andrei Sakharov's birthday brings to mind the one occasion on which I met him. It was in that great yellow mansion on Leninskiy Prospekt in Moscow that houses the Soviet Academy of Sciences, during the period when Academician Keldysh presided over the Academy. Keldysh and I were the receiving line for a large reception, and a substantial number of Academicians were present. I had brought with me a delegation of members of our Academy of whom, at this moment, I recall Lew Branscomb, Harvey Brooks, Ed Ginzton, Gordon MacDonald, Herb Friedman, and George Dantzig. A door opened and Sakharov crossed the transom, flanked by two large men. Keldysh left the line and brought Sakharov to me. We exchanged a few pleasantries and I then reminded him that although he had been elected three months earlier, we had not yet received his formal acceptance of election as a Foreign Associate of our Academy. He smiled and graciously indicated that, since his reply would reside permanently in our archives, he was still laboring to assure that it would be in grammatically perfect, idiomatic English. Somewhat incredulous, I assured him that it would be welcome, even in Russian. We again shook hands; he walked back to the door from which he had entered, and left—again flanked by the two huskies. Later that day I learned that only on the previous day had their Academy voted to agree that he and two other Soviet scientists should be permitted to accept our election in recognition of his enormous contributions to physics. I was unable to catch the eye of any of the Americans present and signal someone to intercept him on the way out. Our meeting was limited to those two formal minutes on the receiving line! Six months later, when we learned that he was in grave danger, I formally threatened to close down the exchange program should anything happen to him—a threat that we have never repeated. Later Sakharov wrote to thank me and stated unequivocally that our communication prevented his arrest at the time.

Sakharov represents a complex mixture of forces. His well-known positions with respect to arms control and the prevention of nuclear war, his championship of human rights, his efforts to reform the socialist system of the USSR through honest criticism, and his consummate humanism have won the hearts of many of his countrymen and attracted legions of admirers around the world. Regrettably, the attribute seldom singled out for praise is his devout patriotism.

To be sure, the system he seeks to reform does not, cannot, perceive his criticism as genuine patriotism. More is the pity. The point, however, for the rest of us is embodied in the words of another giant of the twentieth century.

In the corner of the garden that surrounds our Academy, there now stands a noble statue. Engraved in its granite base are these words:

> As long as I have any choice in the matter, I shall live in a country where civil liberty, tolerance, and equality of all citizens before the law prevail.
>
> —Albert Einstein

Einstein had no choice but to flee from persecution while he still could. Sakharov chose to battle for liberty, tolerance, and equality in his beloved native land, for years protected by his exalted status as one of his country's ranking physicists. Cynics might say that Sakharov had no choice because he realized he would not be allowed to leave the USSR. That seems pointless to me, for it is clear that Sakharov had the choice of remaining silent. And, as far as I know, it remains unclear to this day whether he would leave if he could—the course I have repeatedly urged to the highest levels of his government. I would not take it amiss if he were exiled and placed on some outgoing aircraft.

His brand of patriotism is rare in today's world. And it is that very singleness of purpose and selfless devotion to cause which excites the admiration—and the envy—of those of us who aspire to bring some small change into our respective societies.

We watch our institutions become mired in feckless debate, mindless procedures, and endless argumentation, and we forget to heed the simple warnings of the few great and serious men who seemingly effortlessly reduce the complex to the understandable and transmute the opaque to the transparent.

Those of us who honor Sakharov at celebrations such as this can do no better than to rededicate ourselves to the hard choice of speaking out when we perceive wrong about to be committed and of being honest critics of flawed systems.

I am deeply troubled by some of the things I see today in our country and desperately worried about the way the world beyond our shores is going. This is not the time or the place to elaborate on these concerns; I trust there will be other fora in which I can participate in an advocate's role.

One aspect of Sakharov's patriotism that I find most compelling is his effort to assure that the Soviet government is even-handed in its adherence to the Helsinki Accords. This document, fashioned with so much effort over so long a period, stands as a monument to our collective hope and our collective understanding that nations, like individuals, need principles by which to live—moral guidelines—whether they be secular or spiritual.

The Helsinki Accords had impinged only slightly on my consciousness until about two years ago. Of course I was aware of the great event of 1975, when the Final Act was accepted by the thirty-five signatory nations, and I was troubled by the intense efforts of some countries to avoid applying the provisions equally, but the document itself had not truly become part of my intellectual armamentarium. The turning point in that state of affairs came in February

1980 when, as leader of the U.S. delegation to the "Scientific Forum" sponsored by the Conference on Security and Cooperation in Europe (the thirty-five nations that signed the Helsinki Accords), I found myself studying the text like a schoolboy. Naturally enough, what I learned by that study was already well known to people such as Sakharov and the Helsinki Watchers in all of eastern and western Europe.

At the "Scientific Forum" we, the West, struck one small blow for scientific and intellectual freedom. To appreciate the pain of the process, you must understand that, by the rules of the CSCE, all final documents require unanimous consent; each national delegation has a veto concerning every individual word. Nevertheless, we managed to have the following words included in the final document of the meeting:

> It is furthermore considered necessary to state that respect for human rights and fundamental freedoms by all States represents one of the foundations for a significant improvement in their mutual relations, and in international scientific cooperation at all levels.

Professor Amaldi was a tower of strength in the painful negotiations that achieved that small victory. The Soviet Union had successfully resisted any reference to human rights in the report of the Belgrade Conference. The definition and explanations of human rights and fundamental freedoms appear in one "basket" of the Helsinki Accords; the endorsement of international scientific cooperation in another. Soviet diplomats had consistently argued that each basket stands on its own—without linkage. Our seemingly innocuous little sentence was a triumph in that it not merely acknowledged linkage, it urged attention to that linkage and the Soviets agreed to it. Incidentally, it was at that forum that some forty-nine of the scientists officially present, from twenty countries, signed a telegram to Mr. Brezhnev, pleading for freedom of choice for Sakharov and his family, by then already in Gorky.

At that forum we also attempted in a very small way to utilize the process of scientific exchange to improve the circumstances of scientists in the Eastern Bloc countries. Our report observed that:

Since the signing of the Final Act of the CSCE, there has been a significant expansion of international cooperation in research and training and in the exchange of information. Progress, however, has been greater in some areas than in others. It is observed that the present state of international scientific cooperation still requires improvements in various respects. Such improvements should be achieved bilaterally and multilaterally at governmental and nongovernmental levels through intergovernmental and other agreements, international programmes and cooperative projects, and by providing equitable opportunities for scientific research and for wider communication and travel necessary for professional purposes.

The word "equitable" was mine. It does not translate exactly into other languages but we all agreed to "fair, just, and reasonable" for translation purposes. The Soviet delegation resisted until midnight of the final day of the forum.

We have yet to assess the potency of either paragraph in a real situation. But never again can we be told that human rights are exclusively the internal affair of the Soviet Union, irrelevant to the conduct of scientific exchange. They have solemnly agreed otherwise.

Communication and cooperation among scientists is a tradition of five centuries; indeed it is the very essence of science. Because, without communication, science is essentially pointless, science really knows no international boundaries. Scientists from different countries meet each other in the printed pages of scientific journals, at international conferences, and in each other's laboratories. They share a common culture and set of values, even a common scientific language; no natural barriers need impede their communication.

We all agree that a new piece of information, a new understanding, gained *anywhere* benefits mankind *everywhere*, that no nation has a monopoly of talent, that all can benefit from the work of each of the others. Thus the scientific community really is transnational— a single world community. There are barriers—but only those imposed by bureaucrats.

When, in the 1950s, Soviet scientists began once again to attend international scientific congresses, they were warmly welcomed despite the so-called "interpreters" who accompanied them—surely their KGB companions pained our Soviet colleagues more than they pained us. And happily, with time, that practice disappeared. But why do distinguished Soviet scientists still accept invitations to present papers at international meetings—and then fail to appear, or send some unworthy substitute? We have yet to find acceptable the practice of issuing visas to western scientists just shortly before they depart for the airport; we cannot fathom why laboratories whose research reports appear in the scientific literature are closed to us in person; and we deplore the fact that large, scientifically interesting areas of the Soviet Union—such as Kamchatka, the Arctic, and the permafrost belt—should be closed to our earth scientists and biologists. Our bilateral exchange programs operate on the basis that normally the sending side nominates. But the receiving side is free to propose particularly welcome guests to the sender; we cannot understand why those suggestions are almost never acted upon favorably. And we shall never accept the mutilating censorship of scientific journals before they are deemed fit for the eyes of Soviet scientists.

The Academy and IREX operate the largest, oldest, and most visible individual exchange programs with the nations of Eastern Europe and the USSR. The Soviet program dates back to 1958 and has remained virtually intact through a succession of political and diplomatic crises. Even now it is weathering the storm of budgetary cuts by the new Administration—although it will be reduced sharply.

Last February we suspended our programs of bilateral symposia (on theoretical astrophysics, molecular biology, and experimental psychology) specifically in protest of the internal exile of Andrei Sakharov. Understand that that action was painful and deeply repugnant to our council—as Mannie Piore will attest. It was the smallest clear signal of the depth of our distress that we could devise; letters and cables of protest had for too long fallen on unhearing ears.

Yet we were criticized by some for suspending our small program of Soviet bilateral symposia (not our individual exchanges) on the ground that we were deliberately reducing the very type of exchange we consider most essential to scientific progress; on the ground that we were punishing the Soviet scientific community, which, itself, has no control over what happens to such people as Yuri Orlov, Sergei Kovalev, Anatoly Shcharansky, or Andrei Sakharov; and, not quite consistently with the latter argument, on the ground that the cause of peace is deflected or damaged when scientists are prohibited from meeting; and, finally, on the ground that we were not being even-handed, not also cutting exchanges with Argentina or Uruguay or Korea for their violations of human rights.

Our response has been that the suspension of bilateral symposia rests on the fact that the Soviets have for years insisted on bilateralism to the exclusion of virtually all other modes of interchange. We are, in effect, forced to meet on their terms and conditions; therefore, in order to send a message that is loud and clear, we must do so in their chosen environment. And we lack similar opportunity in those countries with which we have no such agreements.

Of course, we have no desire to punish the Soviet scientific community and we certainly agree to the presumption that the individual members of that community are innocent of the acts of indecency commited by an arm of their government with which they have no contact and on which they have no influence. But if we agree to that, we should have a hard time in accepting the notion that two astrophysicists talking their brand of science at a quiet meeting, somewhere, have any more influence in the Kremlin or in the White House concerning the outcome of SALT II, or the invasion of Afghanistan, or the state of play in Poland, or Cuba, than they have on the Soviet government's respect for human rights. If they were unable to converse, the loss would be to science and, quite probably, to the morale of the already somewhat isolated Soviet scientific community, but not to these other, larger causes for which we had hoped so much. It has taken me all these years to acknowledge to myself that the loss to the cause of peace from loss of these innocent meetings would be very, very small indeed. Please understand that I hold in an entirely separate category those meetings, bilateral or multilateral, at which knowledgeable scientists directly address constructive approaches to arms control.

It is also clear that we do not suspend bilateral agreements that do not exist. In Hamburg, our Soviet colleagues complained privately that the West was singling out the Soviet Union for its treatment of dissidents and not paying very much attention to the Uruguays and

Ugandas of the world. Our collective response was that we all do expect more of the nation that prides itself on being "the other scientific superpower." It is precisely because their scientific attainments are comparable to ours that we can expect them to conduct themselves generally by the same standards that we set for ourselves. We also noted that the scientific class in the Soviet Union enjoys relative privileges vastly greater than those accorded to any other scientific community, ours included. One of these privileges, incidentally, is travel to international meetings and exchange visits—a reward which, one must wryly acknowledge, is linked directly to political conformity. Those who come here very occasionally either conform or have remained invisible to the political system; must we conclude that regular visitors not only conform but must be, in some manner, particularly useful to that political system? I loathe thinking so—but cannot avoid the thought.

Today, our relationship with the Soviet Academy continues to be dominated by considerations of human rights. In August 1980 our Academy Council reaffirmed its position to continue with the voluntary exchange of individuals, but to forgo our handful of bilateral symposia, and to foster multilateral scientific meetings involving Soviet scientists. It is difficult to appraise the consequences of that action. There has been, however, a series of repugnant actions involving scientists whose only offense has been to seek to monitor the behavior of their government with respect to the terms of the Helsinki Accords or to file application to emigrate. In what seemed a deliberately flaunting action, Victor Brailovsky was jailed just as the Madrid meeting of the CSCE opened. And Sakharov languishes in Gorky, ever more isolated from us and from fellow Soviet scientists—with little indication that the world outpouring on his behalf has done anything to mitigate his circumstances.

I believed in and worked at détente: I obtained the funds to double our exchange program in the early '70s, spent much of two years negotiating creation of the International Institute for Applied Systems Analysis (IIASA) outside Vienna, and personally persuaded the Soviets to join in designing the common docking mechanism which made possible the Soyuz-Apollo link-up. But, last year, I was also in full concurrence with those actions of our council. And ever since, I have shared that sensation which Robert Oppenheimer must have felt when he said that "at last physics has known sin."

Deliberately to limit communication between members of the scientific community is a moral sin. However, although perhaps an ugly precedent, our action was and is reversible. Even a small favorable signal concerning Sakharov would suffice to reverse our action—which was, in any case, deliberately much more significant for its symbolism than for its magnitude.

No sign has yet come from the Soviet authorities to suggest that the attitude of the world community will temper the practices to which we take such grave exception. And they continue in other ways to violate international mores. Two weeks ago, the Soviet plasma physicist who served as the full-time resident Secretary of the Council of the International Institute for Applied Systems Analysis was revealed to be engaged in industrial espionage in the North Sea. He resigned and returned home. He had not abused his position at the institute but had used it as a "cover." They are hard to live with!

Thus we still confront an unpalatable dilemma. As matters stand, we somewhat limit organized interaction in bilateral meetings sponsored by the Academy, in that small sense "punishing the innocent." But to lift the embargo would be to turn our backs on the reasons for its institution in the first instance—none of which have been ameliorated. We stand pat.

Since the days of the forum, I have also been privileged to address the CSCE in full assembly in Madrid. On that occasion, my remarks were directed to the conditions of intellectual freedom that must prevail if science is to flourish. For those of you who have not studied the Helsinki Final Act, let me assure you that intellectual freedom is indeed one of the freedoms the Act seeks to preserve, encourage, and expand.

Sakharov himself has said:

> . . . Intellectual freedom is essential to human society—freedom to obtain and distribute information, freedom for open-minded debate, and freedom from pressure by officialdom and prejudices. Such a trinity of freedom of thought is the only guarantee against an infection of people by mass myths. . . .

Freedom of thought is the only guarantee of the feasibility of a scientific democratic approach to politics, economy, and culture.

In Madrid, last December, speaking as an American delegate and also as representative of the western community of scientists, I said:

We perceive no essential distinctions between pursuit of truth about the nature of man or of the physical universe and pursuit of truth about the human condition in the societies in which we live. We will continue to speak out for those whose rights have been denied, for the cost of silence is the abandonment of human rights, and that is a price we will not pay.

The patriotism of Andrei Sakharov—to his native land and to the entire world—is a noble beacon for us all. May that light continue to shine, penetrating the dark corners of repression and the gloom of despair. His years have spanned much of this turbulent century, the principles he so shiningly represents are timeless.

We wish him well.

III

ANDREI SAKHAROV

The Responsibility of Scientists

The following essay was written by Andrei Sakharov for delivery to the International Conference in Honor of Andrei Sakharov sponsored by the New York Academy of Sciences, the American Institute of Physics, and the American Physical Society held May 1–2, 1981, at Rockefeller University, New York. English translation courtesy of Khronika Press.

Because of the international nature of our profession, scientists form the one real worldwide community which exists today. There is no doubt about this with respect to the substance of science: Schrödinger's equation and the formula $E = mc^2$ are equally valid on all continents. But the integration of the scientific community has inevitably progressed beyond narrow professional interests and now embraces a broad range of universal issues, including ethical questions. And I believe this trend should and will continue.

Scientists, engineers, and other specialists derive from their professional knowledge and the advantages of their occupations a broad and deep understanding of the potential benefits—but also the risks—entailed in the application of science and technology. They also develop an awareness of the positive and negative tendencies of progress generally, and its possible consequences.

Colossal opportunities exist for the application of recent advances in physics, chemistry, and biochemistry; technology and engineering; computer science; medicine and genetics; physiology and hygiene; microbiology (including industrial microbiology); industrial and agricultural management techniques; psychology; and other exact and social sciences. And we can anticipate more achievements to come. We all share the responsibility to work for the full realization of the results of scientific research in a world where most people's lives have become more difficult, where so many are threatened by hunger, premature illness, and untimely death.

But scientists and scholars cannot fail to think about the dangers stemming from uncontrolled progress, from unregulated industrial development, and especially from military applications of scientific achievements. There has been public discussion of topics related to scientific progress: nuclear power; the population explosion; genetic engineering; regulation of industry to protect the environment; protection of air quality, of flora and fauna, and of rivers, lakes, seas, and oceans; the impact of mass media. Unfortunately, despite the urgent and serious nature of the issues at stake, such discussions are often uninformed, prejudiced, or politicized, and sometimes simply dishonest. Experts, therefore, are under an obligation to subject these problems to unbiased and searching examination, making all socially significant information available to the public in direct, firsthand form, and not just in filtered versions. The discussion of nuclear power, a subject of prime importance, is an instructive example. I have expressed elsewhere my opinion that the dangers of nuclear power have been exaggerated in the West, and that such distortion is harmful.

With some important exceptions (primarily affecting totalitarian countries), scientists are not only better informed than the average person, but also strive for and enjoy more independence and freedom. Freedom, however, always entails responsibility. Scientists and other experts already influence or have the capacity to influence public opinion and their governments. (That influence should not be exaggerated, but it is substantial.) My view of the situation of scientists in the contemporary world has convinced me that they have special professional and social responsibilities. It is often difficult to separate one from the other—the communication of information, the popularization of scientific knowledge, and the publication of endorsements or warnings are examples of activities with both professsional and social aspects.

Similar complications arise when scientists become involved in questions of disarmament: in developing strategy for or participating in international negotiations; in advancing proposals or issuing appeals to governments or to the public; and in alerting them to dangers. Disarmament is a separate, critically important issue which requires a profound, thorough, and scientifically daring approach. I realize that more detailed treatment is needed, but now I will simply outline a few ideas. I consider disarmament necessary and possible only on the basis of strategic parity. Additional agreements covering all kinds of weapons of mass destruction are needed. After strategic parity in conventional arms has been achieved, a parity which takes account of all the political, psychological, and geographical factors involved, and if totalitarian expansion is brought to an end, then agreements should be reached prohibiting the first use of nuclear weapons, and later, banning such weapons.

Another subject which is closely connected to questions of peace, trust, and understanding among countries is the international defense of human rights. Freedom of opinion, freedom to exchange information, and freedom of movement are necessary for true accountability of the authorities, which in turn prevents abuses of power in domestic and international matters. I believe that such accountability would make impossible tragic mistakes like the Soviet invasion of Afghanistan and would inhibit manifestations of an expansionist foreign policy and acts of internal repression.

The unrestricted sale of newspapers, magazines, and books published abroad would be a major step toward effective freedom of information in totalitarian countries. Perhaps even more significant would be the abolition of censorship which should concern first of all the scientists and intelligentsia of totalitarian countries. It is important to demand a halt to jamming of foreign broadcasts which deprives millions of access to the uncensored information needed to form an independent judgment of events. (Jamming was resumed in the USSR in August 1980 after a seven-year interval.)

I am convinced that support of Amnesty International's call for a general, worldwide amnesty for prisoners of conscience is of special importance. The political amnesties proclaimed by a number of countries in recent years have helped to improve the atmosphere. An amnesty for prisoners of conscience in the USSR, in Eastern Europe, and in all other countries where political prisoners or prisoners of conscience are detained would not only be of major humanitarian significance but could also enhance international confidence and security.

The worldwide character of the scientific community assumes particular importance when dealing with such problems. By its international defense of persecuted scientists and of all persons whose rights have been violated, the scientific community confirms its international mandate which is so essential for successful scientific work and for service to society.

Western scientists are familiar with the names of many Soviet colleagues who have been subjected to unlawful repressions. (I shall confine my discussion to the Soviet Union, since I am better informed about it, but serious human rights violations occur in other countries, including those of Eastern Europe.) The individuals I mention have neither advocated nor used violence, since they consider publicity the only acceptable, effective, and nonpernicious way of defending human rights. Thus, they are all prisoners of conscience as defined by Amnesty International. Their stories have much else in common. Their trials were conducted in flagrant violation of statutory procedures and in defiance of elementary common sense. My friend Sergei Kovalev was convicted in 1975 in the absence of the defendant and counsel, that is, with no possibility whatsoever for a defense. He was sentenced to seven years' labor camp and three years' internal exile for anti-Soviet agitation and propaganda allegedly contained in the samizdat news magazine *A Chronicle of Current Events*, but there was no examination of the substance of the charge.

Comparable breaches of law marked the trials of Yuri Orlov, the founder of the Moscow Helsinki Group, and of other members of the Helsinki Groups and associated committees: Victor Nekipelov, Leonard Ternovsky, Mykola Rudenko, Alexander Podrabinek (and his brother Kirill), Gleb Yakunin, Vladimir Slepak, Malva Landa,

Robert Nazarian, Eduard Arutyunian, Vyacheslav Bakhmin, Oles Berdnik, Oksana Meshko, Mykola Matusevich and his wife, and Miroslav Marinovich. Tatiana Osipova, Irina Grivnina, and Felix Serebrov have been imprisoned pending trial. Yuri Orlov's lawyer missed part of the trial proceedings when he was forcibly detained in chambers adjoining the courtroom. Orlov's wife was frisked in a crude way and her clothing ripped during a search for written notes or a tape recorder, all from fear that the court's grotesque secrets might be revealed.

In the labor camps, prisoners of conscience suffer cruel treatment: arbitrary confinement in punishment cells; torture by cold and hunger; infrequent family visits subject to capricious cancellation; and similar restrictions on correspondence.

They share all the rigors of the Soviet penal regimen for common criminals while suffering the added strain of pressure to "embark on the path of reform," i.e., to renounce their beliefs. I would like to remind you that not once has any international organization, such as the Red Cross or a lawyers' association, been able to visit Soviet labor camps.

Political prisoners are often rearrested, and monstrous sentences imposed. Ornithologist Mart Niklus, poet Vasily Stus, physics teacher Oleksei Tikhy, lawyer Levko Lukyanenko, philologists Viktoras Petkus and Balys Gajauskas have all received sentences of ten years' labor camp and five years' internal exile as recidivists. A new trial is expected for Paruir Airikian, who is still in labor camp. Within the last few days I have been shocked by the fifth(!) arrest of my friend Anatoly Marchenko, a worker and author of two excellent and important books: *My Testimony* and *From Tarusa to Siberia*. Imprisoned religious believers include Rostislav Galetsky, Bishop Nikolai Goretoi, Alexander Ogorodnikov, and Boris Perchatkin. Imprisoned workers include Yuri Grimm and Mikhail Kukobaka. Alexei Murzhenko and Yuri Fedorov are still imprisoned. I shall name only a few scientists deprived of their freedom; many others could be added to the list: Anatoly Shcharansky, the young computer scientist now famous around the world; mathematicians Tatiana Velikanova, Alexander Lavut, Alexander Bolonkin, and Vazif Meilanov; computer scientist Victor Brailovsky; economist Ida Nudel; engineers Reshat Dzhemilev and Antanas Terleckas; physicists Rolan Kadiyev, Iosif Zisels, and Iosif

Dyadkin; chemists Valery Abramkin and Juri Kukk; philologists Igor
Ogurtsov and Mustafa Dzhemilev; and Vladimir Balakhonov.

A common violation of human rights, and one that especially af-
fects scientists, is denial of permission to emigrate. The names of
many "refuseniks" are known to the West.

I was banished without a trial to Gorky more than a year ago and
placed under a regimen of almost total isolation. A few day ago the
KGB stole my manuscripts and notebooks which contained extracts
from scientific books and journals. This is a new attempt to deprive
me of any opportunity for intellectual activity, even in my solitude,
and to rob me of my memory. For more than three years Elizaveta
Alekseyeva, my son's fiancée, has been arbitrarily prevented from
leaving the Soviet Union. I have mentioned my own situation because of the absence of any legal basis for the actions taken and because the detention of Elizaveta is undisguised blackmail directed
against me. She is a hostage of the state.

I appeal to scientists everywhere to defend those who have been
repressed. I believe that in order to protect innocent persons it is
permissible and, in many cases, necessary to adopt extraordinary
measures such as an interruption of scientific contacts or other types
of boycotts. I urge the use, as well, of all the possibilities of publicity
and of diplomacy. In addressing the Soviet leaders, it is important to
take into account that they do not know about—and probably do not
want to know about—most letters and appeals directed to them.
Therefore, personal interventions by Western officials who meet with
their Soviet counterparts have particular significance. Western scientists should use their influence to press for such interventions.

I hope that carefully thought out and organized actions in defense
of victims of repression will ease their lot and add strength, authority,
and energy to the international scientific community.

I have titled this letter "The Responsibility of Scientists." Tatiana
Velikanova, Yuri Orlov, Sergei Kovalev, and many others have decided this question for themselves by taking the path of active, self-
sacrificing struggle for human rights and for an open society. Their
sacrifices are enormous, but they are not in vain. These individuals
are improving the ethical image of our world.

Many of their colleagues who live in totalitarian countries but who have not found within themselves the strength for such struggle do try to fulfill honestly their professional responsibilities. It is, in fact, essential to work at one's profession. But has not the time come for those scientists, who often exhibit their perception and nonconformity when with close friends, to demonstrate their sense of responsibility in some fashion which has more social significance, and to take a more public stand, at least on issues such as the defense of their persecuted colleagues and control over the faithful execution of domestic laws and the performance of international obligations? Every true scientist should undoubtedly muster sufficient courage and integrity to resist the temptation and the habit of conformity. Unfortunately, we are familiar with too many counterexamples in the Soviet Union, sometimes using the excuse of protecting one's laboratory or institute (usually just a pretext), sometimes for the sake of one's career, sometimes for the sake of foreign travel (a major lure in a closed country such as ours). And was it not shameful for Yuri Orlov's colleagues to expel him secretly from the Armenian Academy of Sciences while other colleagues in the USSR Academy of Sciences shut their eyes to the expulsion and also to his physical condition? (He is close to death.) Many active and passive accomplices in such affairs may themselves someday attract the growing appetite of Moloch. Nothing good can come of this. Better to avert it.

Western scientists face no threat of prison or labor camp for public stands; they cannot be bribed by an offer of foreign travel to forsake such activity. But this in no way diminishes their responsibility. Some Western intellectuals warn against social involvement as a form of politics. But I am not speaking about a struggle for power—it is not politics. It is a struggle to preserve peace and those ethical values which have been developed as our civilization evolved. By their example and by their fate, prisoners of conscience affirm that the defense of justice, the international defense of individual victims of violence, the defense of mankind's lasting interests are the responsibility of every scientist.

Gorky, March 24, 1981

P.S. After this letter was written, I received word of the tragic death of Juri Kukk in a labor camp. Tatyana Osipova was sentenced to five years' labor camp and five years' internal exile on April 2.

ANDREI SAKHAROV

Open Letter to Anatoly Aleksandrov, President of the USSR Academy of Sciences

Andrei Sakharov sent the following open letter on October 20, 1980, to Anatoly Aleksandrov, president of the USSR Academy of Sciences. The letter was released by Sakharov's wife, Elena Bonner, in Moscow on November 25.

Dear Anatoly Petrovich,

The immediate cause of this letter is the recently received transcript of your April 15 conversation about my case with the president of the New York Academy of Sciences, Joel Lebowitz. That aside, I consider it important to state my position on questions of principle and on the actions taken by government bodies in my case, to respond to certain public accusations, and to discuss the stand taken by my colleagues in the USSR and, in particular, by the Academy of Sciences and its directors.

For two decades I worked as a scientist in the military-industrial complex and then, for more than twelve years, I have joined those persons engaged in a nonviolent struggle for human rights and the rule of law. My life has forced me to devote particular attention to

The translation of this letter was done by Richard Lourie. It appears here in slightly revised form.

questions of war and peace, international security, international trust and disarmament, and their links to human rights and open societies. As my ideas evolved, they often proved to be unorthodox, at odds with the official line and with my own earlier opinions. My life, my goals, and my ideals have changed radically.

Even much earlier, I reached the conclusion that despite our people's passionate will to peace and the government leaders' unquestionable desire to avoid a major war, our foreign policy has often been dominated by an extremely dangerous geopolitical strategy of force and expansion, and by a striving to subdue and destabilize potential enemies. But in "destabilizing" an enemy we destabilize as well the world in which we live. As early as 1955, I found out that our Near East policy was being reversed in order to gain leverage over the West's oil supplies. This shift has brought great calamities to the peoples of that region—to the Arabs, Israel and Lebanon—and has aggravated the worldwide energy crisis. As Soviet military capabilities have increased, similar policies have proliferated and have become more dangerous, destroying with one hand what the other hand is building. Afghanistan is the latest and most tragic example of the harm done by this expansionist mentality.

I am convinced that the prevention of thermonuclear war is our most important problem and must take absolute priority over all other issues. The resolution of that problem involves politics, economics, the creation of international trust among open societies, the unconditional observance of fundamental civil and political rights, and disarmament.

Disarmament, especially nuclear disarmament, is mankind's most important task. Genuine disarmament is possible, in my view, only if it proceeds from a strategic balance of power. I support SALT II as a satisfactory embodiment of this principle and as a prerequisite to SALT III and other subsequent agreements. I favor an agreement repudiating the first use of nuclear weapons if strategic parity in conventional weapons is achieved. I favor an all-embracing agreement on chemical and bacteriological weapons; the recently reported catastrophe in Sverdlovsk underlines the urgency of this. I would condemn an attempt by the West to achieve substantial strategic superiority over the USSR as extremely dangerous. But I am also concerned by the militarization of the USSR and the Soviet destruction of strategic equilibrium in Europe and other parts of Asia and Africa and by Soviet dictates and demagoguery there.

I oppose international terrorism, which undermines peace no matter what the terrorists' goals. States striving for stability in the world should not support terrorism under any circumstances.

A most important concept which in time became the cornerstone of my position is the indissoluble bond between international security and trust on the one hand, and respect for human rights and an open society on the other. That concept was incorporated in the Final Act of the Helsinki Conference but the words have not been turned into deeds, particularly in the USSR and the countries of Eastern Europe. I have discovered the massive and cynical nature of the violations in the Soviet Union of fundamental civil and political rights, including: freedom of opinion and of information; freedom to choose one's country of residence (i.e. to emigrate and to return) and one's domicile within a country; the right to an impartial trial and to a defense; and freedom of religion. A society which fails to respect these rights is a "closed" society, potentially dangerous to mankind, and doomed to degradation. I have met people who are using publicity in their struggle for human rights, rejecting violence as a matter of principle. They have been cruelly persecuted by the authorities. I have been an eyewitness to unjust trials. I have seen the brazen, crude actions of the KGB. I have learned about terrible conditions in places of detention. I have become one of those people you have called an "alien clique" and even accused of treason. They are my friends and they represent the shining strength of our people.

I learned of the struggle to liberate prisoners of conscience throughout the world and that important goal became one of my close concerns. I support Amnesty International's campaign for the worldwide abolition of capital punishment and I have made several appeals for the abolition of the death penalty in our country.

I have taken a fresh look at the economic difficulties and food shortages in the USSR, at the privileges of the bureaucratic and Party elite, at the stagnation of our industry, at the menacing signs of the bureaucracy perverting and deadening the life of the entire country, at the general indifference to work done for a faceless state (nobody could care less), at corruption and improper influence, at the compulsory hypocrisy which cripples human beings, at alcoholism, at censorship and the brazen lying of the press, at the insane destruction of the environment, the soil, air, forests, rivers, and lakes. The necessity for profound economic and social reforms in the USSR is obvious, but attempts to carry them out encounter the resistance of the ruling bureaucracy and everything goes on as before, with the same worn-out slogans. Occasionally something new is tried, but successes are rare. Meanwhile the military-industrial complex and the KGB are gaining in strength, threatening the stability of the entire world; and super-militarization is eating up all our resources.

My ideal is an open pluralistic society which safeguards fundamental civil and political rights, a society with a mixed economy which would permit scientifically regulated, balanced progress. I have expressed the view that such a society ought to come about as a result of the peaceful convergence of the socialist and capitalist systems and that this is the main condition for saving the world from thermonuclear catastrophe.

The era of the Stalin regime's monstrous crimes represents half of our country's history. Although Stalin's actions have been officially condemned, the specific crimes and the scope of repressions under Stalin are carefully hidden and those who expose them are prosecuted for alleged slander. The terror and famine accompanying collectivization, Kirov's murder and the destruction of the cultural, civil, military, and party cadres, the genocide occurring during the resettlement of "punished" peoples, the penal labor camps and the deaths of many millions there, the flirtation with Hitler which turned into a national tragedy, the repression of prisoners of war, the laws against workers, the murder of Mikhoels and the resurgence of official anti-Semitism, all these evils should be completely disclosed. A people without historical memory is doomed to degradation. You used to share this point of view to some degree and I hope that your position has not changed.

I have expressed my ideas in a series of articles, speeches, and interviews which have appeared since 1968. Instead of engaging in serious discussion, official propaganda has deliberately distorted my position. It has been caricatured, reviled, and slandered. I have experienced increasing persecution, threats directed against me and especially against those close to me, and, finally, deportation without trial.

My earliest attempts to speak frankly met with opposition. On November 22, 1955, the day of the triumphant and tragic testing of a thermonuclear weapon, and even before the bodies of the dead victims had been buried, there occurred a clash between myself and Marshal Nedelin. On July 10, 1961, I got into an argument in your presence with General Secretary Nikita Khrushchev.

And still I succeeded (the Minister of the Medium Machine Building Industry, E. P. Slavsky, can confirm this) in being one of the initiators of the Moscow Treaty Banning Nuclear Weapon Tests in the Atmosphere, in Outer Space, and Under Water, which was the first and still the most generally accepted step on the difficult path to averting the nuclear threat.

In 1975 I was awarded the Nobel Peace Prize, the only Soviet citizen ever so honored. In 1980 I was in Gorky while you, the president of the Soviet Academy of Sciences, talked with the president of the New York Academy of Sciences, who had flown especially from the United States to meet with you. And what was your reply to him? Unfortunately, you spoke in the spirit of the disgraceful statement of the forty academicians of 1973 which laid the groundwork for my denunciation in the press at that time, only you spoke with even greater cynicism and disrespect for the common sense of your listener, our mutual colleague in science.

Yes, I do live in better conditions than those of my friends serving long sentences or awaiting trial, among whom are many colleagues. I will mention only a few: the biologist Sergei Kovalev, the theoretical physicist Yuri Orlov, the mathematicians Tatyana Velikanova and Alexander Lavut, the young computer scientist Anatoly Shcharansky, the physicians Victor Nekipelov and Leonard Ternovsky, the mathematician and computer scientist Alexander Bolonkin (all but the latter I know personally). None of them broke any laws. They neither resorted to nor incited violence. They attempted to realize their goals by use of the written and spoken word. They acted as I have, and our fates cannot be separated. I think it would have been only natural had the Academy of Sciences defended these repressed scientists and not permitted them to be slandered by its president. But my case is different in that here the authorities abandoned even that poor imitation of due process which they have employed in persecuting dissidents in recent years. This is inadmissible both as a precedent and as a relapse. Not a single one of the official institutions charged with executing the law accepted the responsibility for my deportation. You know as well as I do that according to generally accepted legal principles *only a court* can determine a person's guilt, fix the form of punishment and its duration. In all those respects my case is an example of flagrant lawlessness and thus my demand for an open trial is a profoundly serious and principled demand. I am not asking for mercy—I am demanding justice.

You say that I am able to carry on my scientific work in Gorky. Yes, I am working, but it is not for a representative of the Academy of Sciences which has helped organize a one-man *sharashka*[1] for me, to treat that fact as if it were some miracle. Yes, I have a roof over my head (people in Gorky say our apartment used to be a KGB safehouse) and my wife brings from Moscow meat, butter, and cheese, not available in Gorky. But the violation of the law is no less for that. The regimen which has been established for me is absolutely illegal (it does not conform to the Corrective Labor Legislation provisions on exiles). Who determined this regimen—the KGB, the Ministry of Internal Affairs, the Procurator's Office? I do not know and you are not able to answer this question either. There is a policeman at my door around the clock. Anyone attempting to visit me ends up at the police station and experiences serious difficulties. Only after considerable delay, I have learned of attempts by people close to me, such as a physician friend and my eighty-two-year-old aunt, to visit me, and there may be others of which I will never learn. But secretly, without the policeman's knowledge, KGB men enter my apartment, through the window, violating the privacy of my home and creating potential danger for me. Your failure to answer my wife's telegram

[1] A special prison facility for scientists as described in Aleksandr Solzhenitsyn's novel *The First Circle*. —Trans.

concerning this in July was inexcusable. A personal jamming device was installed in my apartment—even before jamming was resumed generally in the USSR. The company is sparing no expense. I am tailed shamelessly, insolently, around the clock. Agents dog my heels everywhere. They peep through my windows, they run ahead of me to the post office to prevent me from making any telephone calls.

In a conversation with Dr. Lebowitz you alluded to my disclosure of state secrets and, at the same time, made unsubstantiated accusations against my friends, claiming that someone had attempted to take abroad secrets received directly from me or through friends. Oddly identifying yourself and the Academy of Sciences with the organs of criminal investigation, you said that "we have detained this person." But legal facts are distinguished from demagoguery and idle gossip. You cited no specific details, nor could you have. In such serious matters unsubstantiated statements are known by another name: slander. With surprising legal flippancy you declared that for my appeals to foreign governments I could be sentenced to five years' imprisonment. Why five years? The maximum sentence under Article 190-1 of the Criminal Code is three years; under Article 70, seven years; and under Article 64, fifteen years or the death penalty. You also mentioned that I could have been assassinated like Kennedy or King. I am not a member of a foreign parliament, and I never appeal to any. But I consider myself obliged to voice my opinion on certain critical questions and to condemn those actions of the USSR which directly contradict its international commitments and international norms. I approve those lawful actions of foreign governments which can help to correct this situation. I have supported the Jackson Amendment and continue to deem it extraordinarily important. This is an amendment to the American law governing trade and is a question of American trade regulation. I appealed to the Indonesian government for an amnesty for their political prisoners. I am accused in the press of praising the coup in Chile because, together with Galich and Maximov, I wrote a letter concerning the fate of Pablo Neruda. Twice I have criticized cruel actions taken against the Kurds in Iraq. Several years ago I appealed for humanity during the siege of the Palestinian refugee camp Tel-Zaatar. In the fall of 1979 I appealed to the government of the People's Republic of China to review the cruel sentence given to a brave dissident who opposed the military action against Vietnam, Wei Jingsheng, and to the government of the Czechoslovak Socialist Republic to review the sentences pronounced on members of Charter 77. I did not support the proposal to boycott the Moscow Olympic Games or the embargo on technology, let alone the grain embargo, before the Soviet invasion of Afghanistan. My position changed when, in my opinion (and in the opinion of 104 countries in the United Nations), there had occurred a dangerous violation of international law and international equilibrium. I then considered the boycott to be for the good of our country. I sent on to the French president Giscard d'Estaing a letter from a group of Crimean Tatar activists and, in my own name, addressed a request to Leonid Brezhnev to end national discrimination against the Crimean Tatars, victims of Stalin's criminal actions in 1944. In October of 1979 I requested Brezhnev to expedite the delivery of food to the starving people of Cambodia. Even after my deportation to Gorky I sent Brezhnev a long letter containing what I considered acceptable proposals for a political settlement of the Afghanistan tragedy. I sent copies of the letter to the heads of the states which are permanent members of the Security Council. In that letter I voiced my opinion that the invasion of Afghanistan was an error having enormous negative consequences for us both in foreign policy and domestically. In particular, I wrote about the strengthening of the repressive organs which could grow out of control. Such are a few of my foreign policy statements in recent years. None of my actions have violated Soviet law. These statements were dictated by my convictions and, in my opinion, did not in any way contradict the interests of our country and its people.

On August 12, 1980, I addressed the vice-president of the Soviet Academy of Sciences, Evgeny Velikhov, and, through him, the Presidium of the Academy of Sciences and you personally, with a request for assistance in a case which has become especially important to me. Here is the story.

The repeated threats directed at my children and grandchildren (beginning with the "visit" of the Black September terrorists in 1973) and other incidents and provocations had compelled us to convince them to emigrate. This was not an easy decision and to this day is felt as a tragedy. Our son's fiancée, Elizaveta Alekseyeva, remained in the USSR and for three years has not been able to leave the country to join the man she loves. She has been subjected to threats and blackmail by the KGB and, although a member of our family, she has not been allowed to visit me in Gorky. Fearing for Elizaveta's life, my wife has been compelled to spend the greater part of her time in Moscow. In point of fact, Liza Alekseyeva has become a hostage. I asked you to intercede in obtaining permission for her to leave the country. In the course of three months the vice-president of the Academy sent no answer whatsoever to my letter or to my repeated telegrams. Only on the evening of October 14 did a telegram arrive stating that "measures are being taken to clarify the possibility of fulfilling your request." It is impossible to understand why this is so complicated if a person never had access to any state secrets. I am beginning to believe that this telegram was no more than a KGB ruse, a delaying tactic. I find it absolutely intolerable that anyone be held hostage on my account. In this case as well I am compelled to turn to my colleagues abroad for support.

You spoke to Dr. Lebowitz about the visit from my colleagues at the Academy of Sciences' Institute of Physics (FIAN) as if that were proof that I have every possibility of doing scientific work. But no matter how important these visits are for me in my isolation, lacking scientific literature and so on, their total dependence on KGB control to choose the visitors and the time of their visit is absolutely inadmissible. Thus, the first visit of FIAN scientists was timed to coincide with the arrival of Dr. Lebowitz so that you could mention that fact during your meeting with him and the second was timed to coincide with the arrival of the secretary of the U.S. National Academy of Sciences. I have worked in FIAN since 1969 and, before that, from 1945 to 1950, and I ought to have the right, on my own and free from the control of the KGB, to choose with whom I shall discuss science. I wrote concerning this unacceptable KGB control to Academician Ginzburg in a letter of September 15, and I requested that he refrain from sending any further delegations of FIAN scientists. Because of the attitude of the Academy of Sciences and the intolerable conditions imposed on my meetings with FIAN colleagues, I am breaking off official scientific relations with Soviet scientific institutions, the Academy of Sciences and FIAN in particular, and I am hereby informing you of that fact.

Prior to the general meeting of the Soviet Academy of Sciences in March 1980, I addressed the Presidium of the Soviet Academy of Sciences with a request to assist my participation in the meeting which is my right and my duty according to the bylaws. I received the following reply: "Your participation in the general meeting is not anticipated." The meaning of those words was graphically underscored by the actions of KGB agents, pistols in hand, who would not allow me onto the Gorky-Moscow train on the evening of March 4, the day before the general meeting, when I was accompanying my mother-in-law and wished to help her with her suitcases. The Presidium of the Soviet Academy of Sciences allowed the KGB to interfere in the affairs of the Academy, formally allowing me to remain a member of the Academy but depriving me of one of the fundamental rights of an Academician.

In sending you this open letter I am hoping that you also will reply openly, presenting your reasoned replies to the questions raised in this letter and especially to the following: Is the leadership of the Soviet Academy of Sciences prepared, in accordance with the wishes of the world scientific community, to undertake an active defense of my violated rights and the rights of other repressed scientists?

Is the leadership of the Soviet Academy of Sciences prepared to demand my immediate return to Moscow and an open trial which will determine my guilt or innocence, and, if I am found guilty, will fix the nature and term of my punishment?

Is the leadership of the Soviet Academy of Sciences prepared, decisively, in deeds and not just words, to defend me from blackmail directed against Elizaveta Alekseyeva, a member of my family, and to aid her to leave the USSR?

The attitude of the Academy of Sciences and its leadership, not only in my case but in the cases of other repressed scientists as well, does not correspond to the traditional understanding of solidarity among scientists. Scientists now bear great responsibility for the fate of the world and this obliges them to remain independent from bureaucratic institutions, especially the secret police, whether known as the FBI or the KGB. I continue to hope even now that the Academy of Sciences will display such independence.

Respectfully,
ANDREI SAKHAROV
Member of the Soviet Academy of
Sciences since 1953

Gorky, October 20, 1980

ANDREI SAKHAROV

A Letter from Exile

I should like to offer some thoughts on problems that have been troubling me and discuss the way they appear to me here in Gorky, a city closed to foreigners, in the depths of the Soviet Union, where I now live under vigilant surveillance by the KGB.

World Problems

In the 1960s and '70s, the Soviet Union, making use of its growing economic and scientific-technological potential, carried out a fundamental reequipping and expansion of its weaponry. There was a substantial increase in both the quality and quantity of the missiles and nuclear weapons developed earlier, and in other new systems of military technology: transport vehicles for the largest land force in the world, the latest word in tanks and aircraft, combat helicopters, fire-control systems, communications, nuclear submarines, fast hydrofoil craft, and many others.

A major change has occurred in the world balance of forces, and this change is intensifying. It is true, of course, that the development of new technology and the growth in numbers of weapons have not been confined to the Soviet Union. This is a mutually stimulating process in virtually all technologically developed countries. In the

Translated by Raymond H. Anderson.

United States, in particular, such developments have perhaps proceeded on a higher scientific-technological level and this, in turn, caused alarm in the Soviet Union.

But in order to assess the situation properly it is imperative to take note of the particular features of the Soviet Union—a closed totalitarian state with a largely militarized economy and bureaucratically centralized control, all of which make the growing might of such a country even more dangerous. In more democratic societies, every step in the field of armaments is subjected to public budgetary and political scrutiny and is carried out under public control. In the Soviet Union, all decisions of this kind are made behind closed doors and the world learns of them only when confronted by *faits accomplis*. Even more ominous is the fact that this situation applies also to the field of foreign policy, involving issues of war and peace.

At the same time that the change in the balance of forces was occurring—though not only because of that change—there was both covert and overt Soviet expansion in key strategic and economic regions of the world. Southeast Asia (where Vietnam was used as a proxy) and Angola (with Cuba as the proxy), Ethiopia, and Yemen are only some of the examples. The invasion of Afghanistan may be a new and more dangerous stage in this expansion. The invasion, which occurred against the background of the tragedy in Tehran, and possibly had some concealed connection with it, exacerbated world tensions and obstructed talks on disarmament and the settlement of other conflicts. In particular, the invasion made it impossible, at the present time, for the United States Congress to ratify the SALT II treaty, which is of such crucial importance to the world.

Several months earlier, the Soviet Union had unleashed at home and had instigated abroad a demagogic campaign against plans by the United States and North Atlantic Treaty Organization allies for an urgently needed modernization of their missile forces in Europe—this at a time when the Soviet Union had already completed such a modernization—and had thwarted (I hope only temporarily) talks on limiting medium-range nuclear missiles. The Vienna negotiations on disarmament in Europe are likewise in a deplorable state, which is also mainly the fault of the Soviet Union.

Despite all that has happened, I feel that the questions of war and peace and disarmament are so crucial that they must be given absolute priority even in the most difficult circumstances. It is imperative that all possible means be used to solve these questions and to lay the groundwork for further progress. Most urgent of all are steps to avert a nuclear war, which is the greatest peril confronting the modern world. The goals of all responsible people in the world coincide in this regard, including, I hope and believe, the Soviet leaders—despite their dangerous expansionist policies, despite their cynicism, dogmatic conceptions and lack of self-confidence which often prevent them from conducting more realistic domestic and foreign policies.

Therefore, I hope that when there is some easing of the present crisis in international relations, caused mainly by the Soviet invasion of Afghanistan, there will be a revival of efforts in regard to SALT II, a technologically progressive treaty that provides the essential foundations for SALT III. I hope, too, that there will be new efforts in regard to medium-range nuclear weapons and tactical weapons as well as a reduction in "ordinary" weaponry in Europe.

Negotiations on disarmament are possible only on the basis of strategic parity. The countries of the West must do everything necessary to maintain this parity or, in some categories, to regain it—not allowing themselves to become victims of blackmail and demagogy as in the campaign against American missiles in Europe.

Of equal urgency is a peaceful settlement of "hot" conflicts. The Soviet invasion of Afghanistan was condemned by 104 nations, but the war continues there and no end is in sight. Economic and political sanctions are extremely important; they can help strengthen the hand of the more responsible, nondogmatic members of the Soviet leadership. In particular, the broadest possible boycott of the Moscow Olympics is necessary. Every spectator or athlete who comes to the Olympics will be giving indirect support to Soviet military policies.

It is vital to demand withdrawal of the Soviet troops in Afghanistan. I hope that withdrawal of the troops will become possible—if not now, perhaps later—on the basis of guaranteed neutrality, with stationing in the country of United Nations troops or units from neutral Moslem countries, which should ease Soviet apprehensions.

The Soviet "conditions" that foreign interference be ended are pure demagogy because there is no such interference. In general, Soviet propaganda is conducted now on a crude "military" level. For example, television broadcasts are showing allegedly captured "American" grenades containing nerve gas. Painted on the grenades in large white letters are the words "Made in U.S.A." All this is obviously intended to head off any similar accusations against operations by the Soviet Army.

The Middle East conflict has been dragging on for decades. The main hope for its settlement is development of the Camp David line so that the difficulties and tragedies of the past will not be carried into the future. It is very important that the Palestinians adopt such a position, that they recognize the existence of Israel, renounce terrorist-guerrilla methods of struggle and refuse to be pawns of those who supply them with guns. There is no evidence yet of any movement in this direction.

Israel, it seems to me, should show more restraint, particularly in regard to settlements in the occupied lands. The Middle East crisis can be solved only in the context of general world problems. Therefore, United States mediation seems to me to be of utmost importance.

In November 1979, I wrote to Leonid I. Brezhnev, chairman of the Presidium of the Supreme Soviet of the USSR, urging him to help assure uninterrupted supplies of food to the starving people in Cambodia, where famine is the consequence of the crimes of the deposed Pol Pot regime and the war. This problem, like that of the refugees from Vietnam, Cambodia, Ethiopia and other countries, remains as critical as ever. Now there is the problem of refugees from Afghanistan and Cuba. Saving people must be put ahead of all military and political considerations and national prestige.

At the moment I write this, the world has just learned of the failure of the American attempt to rescue the hostages in Tehran. I feel that this was a brave and noble effort. It was undertaken only after it had become clear that the Iranian authorities would not free the hostages voluntarily, thereby taking upon themselves responsibility for a crime that flouts the standards of international behavior.

I do not understand how so many technical problems could have arisen. These are no doubt being subjected to intensive scrutiny, not only the technical questions but also the possibility of sabotage. Success of the American mission would have saved the world from a nightmare. No one should condemn the United States because the mission failed. And no one should criticize President Carter for the secrecy in which the mission was organized. It would be my guess that the secrecy, if anything, was insufficient rather than excessive. Personal ambitions had no place in this. Overall, the actions of Mr. Carter in these tragic days win only respect from me.

Success of the American rescue mission would have eliminated a need to impose sanctions against Iran, and it would have been in the interest of the Iranian people themselves. It appears now that sanctions have become inevitable, and it is very important to achieve unity of action by the countries of the West. Solution of the hostage problem is a matter for all countries that consider themselves democratic and strive to preserve the norms of international relations. This is not just an American problem but a worldwide one, as was shown by new acts of terrorism: the successfully resolved incidents in Bogotá and London.

World attention should be paid to many other problems whose burden now rests mainly on the United States—economic and technological aid to developing countries, help to refugees and the starving, and general economic, ideological, and military resistance to the expansion of totalitarianism.

Western unity is one of the main conditions for international security, a unity that will promote resistance and ultimately lead to rapprochement and the convergence of world systems, averting thermonuclear catastrophe.

Western Problems

A totalitarian system conducts its policy through control from a single center—diplomacy, information, and disinformation services inside and outside the country, foreign trade, tourism, scientific-technological exchanges, economic and military assistance to liberation (a word that must sometimes be used with quotation marks) movements, foreign policy of satellite countries, and all kinds of clandestine activities—all these are coordinated from a single center. Special attention must be paid here to clandestine activities, since a person is inclined to forget something if it is not waved in front of his eyes. The West and developing countries are filled with citizens who by reason of their positions are able to promote Soviet influence and expansionist goals.

Some of them are motivated by ideas that at least merit discussion. After all, in the Soviet Union, the ideological epicenter, and in China as well, Communist ideology is not a complete fraud, not a total delusion. It arose from a striving for truth and justice, like other religious, ethical, and philosophical systems. Its weakness, its failure, and its degradation—evident from the very beginning—represent a complex historical, scientific, and psychological phenomenon that requires separate analysis.

There are others among such people who conduct themselves in a "progressive" manner because they consider it profitable, prestigious, or fashionable.

A third category consists of naive, poorly informed, or indifferent people who close their eyes and ears to the bitter truth and eagerly swallow any sweet lie.

Finally, there is the fourth group—people who have been "bought" in the most direct sense of the word, not always with money. These include some political figures, businessmen, a great many writers and journalists, government advisers, and heads of the press and television. Overall, they make up a substantial group of influential people.

(I can't resist telling about an incident involving my wife and me. Two leading and influential American scientists, arriving in the Soviet Union to take part in a conference, were given envelopes containing money for personal expenses. Out of politeness and embarrassment they felt they could not decline the money. So, in a state of confusion, they gave it to us and asked that we pass it on to scientists who had lost their jobs. We don't know how many other such incidents there have been.)

Of course, there are many spies, secret agents, and organizers of sabotage. This is true of nations other than the USSR, but a totalitarian system has special opportunities. In particular, one cannot arbitrarily reject allegations by some writers about connections between the KGB and international terrorism. The absence of direct proof regarding such links and concern about a further worsening of tensions make governments of the West reluctant to look too deep into this sensitive question.

Unity of all its forces is one of the advantages of totalitarianism in its world offensive threatening the pluralist West. What can the West do to counter this challenge? Of course, in historical perspective, in conditions of peaceful and orderly development, the pluralist and free systems are more viable and dynamic. Therefore, the future will follow the path of pluralist and converging scientific-technological progress.

But the world is facing very difficult times and cruel cataclysms if the West and the developing countries trying to find their place in the world do not now show the required firmness, unity, and consistency in resisting the totalitarian challenge. This relates to governments, to the intelligentsia, to businessmen, and to all people. It is important that the common danger be fully understood—everything else will then fall into place. In this regard, I believe in Western man. I have faith in his mind, which is practical and efficient and at the same time aspires to great goals. I have faith in his good intentions and his decisiveness.

Recent months have been under the shadow of the events in Afghanistan and Tehran. The reaction in Europe, at least the initial reaction, has not been as consistent and united as, in my opinion, it should have been. One could read assertions in the press of the following type: Let Carter worry about it. He's the one facing the elections. After all, this is a purely American matter. We have our own problems.

I am unable to judge how widespread are such views. In my opinion, they are very harmful. Recently, the West German author Günter Grass and three other writers issued a statement in this spirit. The Soviet press delights in quoting such statements, forgiving the authors all their past "sins." The anti-Americanism of certain representatives of the West European intelligentsia would be merely amusing, given their show of exaggerated sensitivity, if it were not so harmful.

In truth, Europe has much of which it can be proud. But it is inappropriate for Europe to adopt an arrogant attitude. The tragic nature of our time does not permit this. Europe must fight shoulder to shoulder with the transoceanic democracy, which is Europe's creation and Europe's main hope. A certain lack of unity, of course, is the reverse side of the coin of democratic pluralism, the West's major strength. But this disunity is also caused by the systematic Soviet policy of driving "wedges," a policy that the West has not resisted adequately because of carelessness and blindness. Nonetheless, I feel that in the course of recent crises a positive shift has appeared in the stand of the West and of the developing countries. Only the future will show whether I am right.

Among the important events of recent years is a movement toward greater independence by several European Communist parties—though now the French have apparently beaten a quick retreat. There may be profound consequences if the parties continue to reject dogmatism and unquestioning support of Moscow (in such matters as Afghanistan, for example) and if they accept certain pluralist principles. It would be very important if the European Communist parties joined all democratic forces in supporting the struggle for human rights in the Soviet Union and other Communist countries.

One of the causes of the weakening position of the West is its dependence on oil supplies, a "fatal dependency," as an American leader has remarked. The geopolitics of the Soviet Union is aimed precisely at this weak point. In these circumstances, the West cannot afford to deprive itself of nuclear energy, which gives room for diplomatic and economic "maneuvering." Concern about safety and environmental hazards should have no bearing on the principal issue— to build or not to build nuclear stations—but only on how to build them. The price in terms of polluting the environment is greater from coal and oil than from nuclear energy. Of course, other sources of energy must be developed as an alternative to oil, including coal, despite its ecological shortcomings. Strict conservation must be introduced in the use of electricity and heat. Steps must be taken to provide small cars, good public transportation, insulation of homes, integrated heating systems and, especially, introduction of technological processes in industry to conserve electricity. Western voters

must demand this from government leaders and not allow dema-
gogues to exploit these problems for their selfish ends.

Internal Problems

Defense of human rights has become a worldwide ideology, uniting
on a humane basis peoples of all nationalities and with the most diverse convictions. I have very high regard for them all: for Amnesty
International and its struggle for release of prisoners of conscience,
against torture and the death penalty; for the International League
for Human Rights; and for the rights activists in Eastern Europe,
China, and other countries, where they show great bravery in coping
with cruel repression.

In the Soviet Union, the movement for human rights emerged in
its present form at the end of the '60s with publication of the Chronicle of Current Events, an anonymous underground journal that reports cases of violations in the Soviet Union factually and without
subjective comment. Despite severe repressions, the journal has continued to appear, a total to date of fifty-four issues. In more recent
years, appeals by the Helsinki Group [formed in the Soviet Union to
check on compliance with human rights provisions of the 1975 Helsinki agreement on European security] began to appear.

The human rights movement has no political objectives and its
participants have no desire to gain political power. Their only weapon
is free access to and dissemination of information. It is of vital importance that the movement limit itself to nonviolent methods. Such a
position is logical in a country that has passed through the violence
of every circle of hell. Calls for new revolutionary upheavals or for
intervention would be mad, and a terrible crime in an unstable world
only several steps from the thermonuclear abyss.

Participants in the human rights movement speak out openly for
human rights whenever they learn of violations, and they inform the
people. They have also set themselves the task of correcting the historical record about a society and individual citizens if the truth has
been distorted by official propaganda. They help the families of victims of repression. I am convinced that this is what is needed—a
pure moral movement to plant in people's minds a basis for demo-

cratic and pluralist transformation. This is crucial to the country and essential to all mankind for the sake of peace on earth.

The consciousness of broad masses of the population has been deformed by a number of factors: Decades of totalitarian terror . . . old and new prejudices . . . the lure of a relatively good life after generations of havoc (I mean of course a very limited good life, nothing like the well-being and freedom of workers in the West or the privileged elites in the Soviet Union) . . . the constant need to wheel and deal, to scheme and break the law. The ideology of the Soviet philistine (I have in mind the worst people, but they, unfortunately, are rather frequently found among workers and peasants and throughout the intelligentsia) consists of several uncomplicated ideas:

• Cult of the state, involving, in various combinations, submission to authority, a naive belief that life in the West is worse than in the Soviet Union, gratitude to a "benefactor" government, and, at the same time, fear and hypocrisy.

• Egoistic endeavors to ensure a good life for oneself and one's family, to "live like everyone else" with the help of graft, theft ignored by bosses, and ever-present hypocrisy. Yet there is a desire among better people in this category to achieve a good life through their own labor, by their own hands. Nonetheless, it is still necessary to wheel and deal and to play the hypocrite.

• The idea of nationalist superiority, which takes on a dark, hysterical, and pogromlike form among some Russians, and not only among Russians. One often hears exclamations: "We're wasting our money on these black (or yellow) monkeys! We're feeding parasites!" Or one hears: "The Jews are responsible!"—or the "Russians" or the "Georgians" or the "Chuchmeki," a derogatory term for the peoples of Central Asia.

These are very disturbing symptoms after sixty years of proclaiming "friendship of the peoples."

Officially, Communist ideology is internationalist, but it surreptitiously exploits nationalist prejudices. So far, this has been done with some caution, and I hope that these forces will not be unleashed. After the class hatreds we have suffered, we certainly have no need for a racist-nationalist ideology. I am convinced that this is dangerous and destructive even in its most humane (at first glance) "dissident" manifestations. There are few people who react seriously anymore to slogans about building Communism, although there was a time when, perhaps as a result of a certain misunderstanding, Communist slogans reflected a wish for justice and happiness for all in the world.

But internal propaganda intensively exploits the nationwide tragedy of World War II and the pride that people feel in their active part in historic events of that time. The irony of life is that it was only during the war that the ordinary person felt his importance and his dignity in an inhuman world of terror and humiliation. There is intensive exploitation of the risk of war and the much decried American military bases around our country. Feelings of suspicion are stirred up about schemes of the "imperialists."

A nation that has suffered the horrible losses, cruelties, and destruction of war yearns above all for peace. This is a broad, profound, powerful, and honest feeling. Today, the leaders of the country do not, and cannot, go against this dominant desire of the people. I want to believe that in this regard, the Soviet leaders are sincere, that when peace is involved they are transformed from robots into people.

But even the people's deep wish for peace is exploited, and this is perhaps the cruelest deception of all. The deep yearning for peace is used to justify all the most negative features in our country—economic disorder, excessive militarization, purportedly "defensive" foreign policy measures (whether in Czechoslovakia or Afghanistan), and lack of freedom in our closed society. And those negative features also include the ecological madness, such as the destruction of Lake Baikal, meadows and fields, and the country's fish resources, and the poisoning of our water and air.

The people of our country submit uncomplainingly to all the shortages of meat, butter, and many other products—though they do grumble at home. They put up with the gross social inequality between the elite and the ordinary citizens. They endure the arbitrary behavior and cruelty of local authorities. They know about the beatings and deaths of people in police stations but as a rule keep quiet. They do not speak out—sometimes they even gloat—about the unjust treatment of dissidents. They are silent about any and all foreign policy actions.

A country living for decades under conditions in which all means of production belong to the state is suffering serious economic and social hardship. It cannot grow enough food for its people. It cannot, without the benefit of détente, keep up with the contemporary levels of science and technology.

From the time I wrote "My Country and the World" [published in 1975], the average salary has risen, but the cost of living has evidently risen even higher, because daily life has not improved. The much acclaimed free medical care steadily gets worse. (It is "free" because the wages of most workers are kept so low and because one must pay for expensive medicines.) The situation in education is not much better, especially in the countryside. It is no longer possible to explain away all these problems as a result of the war or occasional mistakes.

There is an urgent need for economic reforms that would increase the independence of enterprises and allow elements of a mixed economy. There is need for more freedom of information, a free and critical press, freedom for people to travel abroad, freedom of emigration, and a free choice of one's place of residence within the country. In the long run, there should probably be a multiparty system and elimination of Party monopoly over all ideological, political, and economic life.

But all of this, even though obvious to most people, remains for the time being nothing but wishful thinking. The dogmatic bureaucrats and the new people replacing them—anonymous and shrewd cynics, moving in the many "corridors of power" of the departments of the Central Committee, the KGB, the ministries, and the provincial and regional Party committees—are pushing the country toward what they consider to be the safest path but that is in reality a path to suicide.

Everything is as it was under the system created by Stalin. The leaders carry on the arms race, concealing it behind talk of their love of peace. They interfere in troubled areas around the world, from Ethiopia to Afghanistan, in order to increase prestige, to strengthen the nation's power, and to ensure that the guns don't get rusty. They round up dissidents, returning the country to the quiet "pre-dissident" period, as my son-in-law, Efrem Yankelevich, has described the situation. [Mr. Yankelevich, who emigrated in 1977, is a researcher at the Massachusetts Institute of Technology.]

In the last ten or fifteen years there has been a worsening of the traditional Russian curse, drunkenness. The government has attempted some timid half-measures—more in word than deed—but it is unable to accomplish much. Alcoholism is a worldwide phenomenon, not wholly a result of conditions in our country. But certain specific factors do play a major role.

Expenditures for drinking reduce surplus purchasing power of the population, but the main point is that an alcoholic poses no threat to the government. Also, drinking is the only real freedom available and the authorities are not foolhardy enough to take this away without giving something in return. There are economic, social, and psychological elements in all this. And the result: Instead of dry wine and good-quality old vodka, the authorities flood the market with cheap and poisonous fortified wine, known as *bormotukha*, which swiftly destroys men, women, and youngsters. As the "quiet" tzar, Aleksei Mikhailovich, said three hundred years ago: "Don't drive the hot-heads away from the taverns."

The people of our country are to some extent confused and intimidated, of course. But there is also a conscious self-deception and an egoistic escape from difficult problems. The slogan "The People and Party Are One," which hangs from every fifth building, consists not entirely of empty words.

But it was from the ranks of the people that the defenders of human rights emerged, standing up against deceit, hypocrisy, and silence, armed only with pens, ready to make sacrifices, yet lacking the stimulus one derives from the certainty of quick success. They had their say. They will not be forgotten. On their side, they have moral force and the logic of historical development. I am convinced also that their activity will continue in one form or another, whatever the size of the movement. What is important is not the arithmetic but the qualitative fact of breaking through the psychological barrier of silence.

But history develops according to its own slow (and tortuous) laws. We are now living through difficult and troubling times—a worsening of international tensions, Soviet expansionism, shameless anti-American, anti-Western, anti-Israeli, anti-Egyptian, and anti-intellectual propaganda, and threats of still greater tension ahead.

Inside the country, these are times of ever greater repression. It is terrible to think that the most honorable and generous people, who have devoted many years to defending others through public protest, have fallen victim to arbitrary repression.

I feel obligated to tell something about a few of them: Tatyana Velikanova, a mathematician and mother of three children, a grandmother, participant in the struggle for human rights for more than twelve years, from the very beginning to the present. Showing no interest in fame, glory or personal gain, sacrificing much in her personal life, she has always been at the center of the battle, committing herself to the fate of hundreds of victims of injustice, speaking out on their behalf, helping them in every way she could, not caring whether their opinions were close to hers or distant. Her only consideration was whether someone had suffered injustice.

I do not reproach those who could not bear up under the many years of strain, those who quit the struggle, or even those who in some way betrayed themselves. But they demonstrate all the more why we should admire the courage of this woman.

Another is Malva Landa, a geologist, an active member of the Helsinki Group, one of the volunteers in the aid fund helping dissident families, and for many years, for decades in fact, a friend of political prisoners and their families, totally dedicated to the idea of justice.

It is the same with Sergei Kovalev, a talented biologist and a deep and penetrating thinker, kind, patient, and strong. We were all devoted to him when he was free. I was often impressed by the deep respect he received from many of his fellow prisoners during six years in a labor camp.

Another is Victor Nekipelov, who gave his utmost to help others in trouble or subjected to injustice, a sensitive poet, a loving father, a brave man.

All the world knows of Prof. Yuri Orlov, the physicist, a courageous man always in the forefront, founder of the Helsinki Group.

The world also knows of Anatoly Shcharansky, falsely accused of espionage in an attempt to intimidate the Jewish emigration movement.

I have deep respect for the talented writer and World War II invalid Mykola Rudenko; Vyacheslav Bakhmin, the courageous and honorable member of the Working Commission on Psychiatric Abuse; Leonard Ternovsky, a radiologist and member of the same commission and also of the Helsinki Committee in Moscow, a remarkably kind and steadfast man.

Repression has been intensified against religious believers and those who defend their cause. Noteworthy among these are the names of the priests Gleb Yakunin and Dmitri Dudko and of Victor Kapitanchuk, Lev Regelson, Alexander Ogorodnikov, and Vladimir Poresh. Also, the names of the church elder Nikolai Goretoi and of eighty-four-year-old Vladimir Shelkov, who recently died in a camp, should be well known.

Mustafa and Reshat Dzhemilev and Rolan Kadiyev, fighters on behalf of the Crimean Tatars, are once again imprisoned.

As I was writing this article, more tragic news reached me: the arrest of Alexander Lavut, a talented mathematician and one of the veterans of the struggle for freedom of information. I have known Lavut for many years. Modest, serious-minded, and good-natured, he never sought to draw attention to himself. But he worked on behalf of many people. Many of them, including me, will miss his kind words and sound advice. All those I have listed have either been sentenced to long terms or are awaiting an illegal trial. All those who are free have the inescapable duty to speak out on their behalf and on behalf of the many others I have not mentioned.

Some Words About Myself

I live in an apartment guarded day and night by a policeman at the entrance. He allows no one to enter but family members, with a few exceptions. There is an old friend who lives in Gorky; the price he pays for associating with us is a summons to the KGB for a talk after each visit. There is a Gorky physicist who has been denied an emigration visa; he also has to go to the KGB after his visits to us. The only other visitors are people judged suitable by the KGB. There is no telephone in the apartment. I am unable to telephone Moscow or Leningrad even from the public telephone bureau at the post office. The call is immediately disconnected at the orders of the KGB agents who always follow me. I receive very little mail, and that consists mainly of letters "reeducating" me or merely cursing me. Oddly enough, I get the same kind of letters from the West. However, I also get some mail from the West with kind words and I am deeply grateful to the senders.

When I accompanied my mother-in-law to the station on her departure for Moscow, KGB agents, pistols in hand, made a show of preventing me from approaching the coach, making it clear that the order forbidding me from going beyond the city limits was not just empty words. A radio-jamming facility has been set up in the apartment building, just for me. In order to listen to the radio, it is necessary for my wife and me to go for walks at night with a transistor receiver. While we are out walking, the KGB agents are in our apartment damaging the typewriter and tape recorder or searching through our papers. I occupy myself with scientific work, but I suffer from a lack of regular contact with colleagues.

At the end of the third month of my stay in Gorky, on the eve of the arrival in Moscow of Western participants in an unofficial scientific seminar, the KGB allowed my colleagues in the Physical Institute of the Academy of Sciences to visit me. It even recommended that they do so. I am very grateful to those who did come. It had been a long time since I had an opportunity to discuss developments in science. During their stay, the police post was moved away from the door and the jamming facility was turned off. But as soon as my colleagues had left, promising that others would come in time, everything returned to the former state. And there was a resumption of the regular summons to the Ministry of Internal Affairs for registration.

In terms of everyday life, my situation is much better than that of my friends sent into exile or, particularly, sentenced to labor camp or prison. But all the measures taken against me have not even a shred of legality. It is part of a harsh, nationwide campaign against dissidents, including the attempt to force me to keep silent and thereby make it easier for repressive action against others.

On January 22, in Moscow, KGB agents forcibly took me to the Deputy Prosecutor General, Aleksandr M. Rekunkov, who informed me that I was being stripped of my awards and would be sent into exile. He produced only the decree of the Supreme Soviet concerning the awards, giving an impression, however, that the decree also called for exile. But that was not so. I still do not know which branch of government or who personally made the decision to have me exiled. All my inquiries meet only with silence. In any event, the decision is illegal and is in violation of the Constitution. In two letters to Rekunkov and in a telegram to the chairman of the KGB, Yuri V. Andropov, I demanded revocation of the illegal exile order and said I was prepared to face an open trial.

In recent months the Soviet press has printed many articles accusing me of "mortal" sins—contempt for the people and their aspirations, slander of the Soviet system, inciting the arms race, groveling before American imperialism, and divulging military secrets. I shall not reply, once again, to those accusations here. In part, this article is my reply. I put forth my position, in very brief form, in my first statement from Gorky. All my activities stem from a desire for a free and worthy destiny for our country and our people and all countries and peoples of the world. I consider the United States the historically determined leader of the movement toward a pluralist and free society, vital to mankind. But I equally respect all peoples for their contributions to our civilization and future society.

In April, the former president of the New York Academy of Sciences, Dr. Joel Lebowitz, visited Moscow. He passed on to the president of the Soviet Academy of Sciences, Anatoly P. Aleksandrov, a petition from American scientists urging that I be freed from exile and be allowed to return to Moscow or, if I should so choose, emigrate to the West. Aleksandrov replied that the exile was for my own welfare because I had been surrounded in Moscow by "dubious characters" through whom there had been a leak of information involving state secrets. That was an outrageous assertion. I have never had dealings with "dubious characters." My friends are honorable and decent people, all of whom are known to the KGB. If state secrets have been betrayed, any guilty persons (including me, first of all) should be put on trial. But the accusation about betrayal of secrets is nothing but slander. The response to the demand that I be allowed to emigrate was also somewhat strange: "We have signed the treaty on nonproliferation of nuclear weapons and we abide by it strictly." As if I were a hydrogen bomb!

I am frequently asked whether I am ready to emigrate. I feel that constant discussion of this matter in the press and many foreign broadcasts is premature, motivated by thirst for sensationalism. I feel that every person has the right to emigrate and, in principle, I do not exclude myself. But I do not regard this as a valid question for me at this time because the decision does not depend upon me anyway.

I regard as just and legal the demands of my foreign colleagues that my exile order be revoked and that I be allowed to return home or go to the West. These are my rights not only as a scientist but also as a human being. I am very grateful for the concern and for the clearly formulated demands.

This article is being taken to Moscow by my wife, my constant helper, who shares my exile and willingly takes upon herself the heavy burdens of traveling back and forth, handling my communications with the outside world, coping with the growing hatred of the KGB. Earlier, she withstood the poison of slander and insinuation, focused more on her than on me. The fact that I am Russian and my wife is half Jewish has proved useful for the internal purposes of the KGB.

Recently, someone appeared at the door of my mother-in-law at 5:30 A.M., describing himself as an officer of the KGB. He warned that if her daughter, meaning my wife, did not stop her trips back and forth from Gorky and stop inciting her husband with anti-Soviet statements, they would take certain measures. Earlier, some of our friends received letters with similar threats against my wife. Every time my wife leaves, I do not know whether she will be allowed to travel without hindrance and to return safely. My wife, although formally not under detention, is in greater danger than I am. I urge those who speak out on my behalf to keep this in mind. It is impossible to foresee what awaits us. Our only protection is the spotlight of public attention on our fate by friends around the world.

May 4, 1980

ANDREI SAKHAROV

Some Thoughts on the Threshold of the Eighties

My isolation in Gorky is of an entirely different nature than appears from the conference. There is no scientific library opposite my home, to which I might have access. Opposite there is only mud and piles of rubbish. I have no contact with scientists in Gorky, not because there are only secret institutes here, but because I am in a state of almost total isolation, deprived of the possibility of meeting anybody at all apart from my wife and two people from Gorky, who obtained permission for this from the KGB. And one visit from my university colleague, also by permission of the KGB. Any others are kept away by a militiaman, on duty around the clock, one meter from the door.

I don't even get to know about the majority of visitors, and they have great trouble. After some time I merely learn of people who are close to me. Our friend and doctor, who traveled from Leningrad, was not admitted, nor was our eighty-two-year-old aunt from Moscow. They do not even admit my son's fiancée, who has lived with us nearly three years, Liza Alekseyeva. The authorities will not allow her out of the country to join the person she loves; she is subjected to persecution, threats of physical and legal reprisals.

To describe my situation, I might add that I have no telephone and it is not possible to make a call from a post office. I am deprived of the medical aid of those doctors who used to treat me; my correspondence is carefully inspected by the KGB, and only a fraction of it reaches me. In the house where I live there is a personal radio-jamming device, which was in operation even before jamming of radio transmissions was resumed in the USSR.

In July my wife found two KGB agents in the flat, who had entered through a window and, without the knowledge of the militiaman on duty, rummaged through my papers and erased tape recordings. Illegal entry like this, the purposes of which may be even more dangerous, has happened before. I have not received a reply to a single one of my letters or telegrams to officials. Two months ago I sent a letter to the vice-president of the Academy of Sciences of the USSR, E. P. Velikov, and would like to hope for a reply.

The Soviet press, Soviet representatives abroad, and some of my Soviet colleagues during foreign missions, in contacts with people in the West who are concerned about my fate, in an attempt to disorganize my defense, assert that I am against détente, have spoken out against SALT, and have even permitted the divulgence of state secrets; they also emphasize the mildness of the measures taken against me. My attitude and open way of life and actions are well known and show how absurd these accusations are.

I have never infringed state secrecy, and any talk of this is slander. I regard thermonuclear war as the main danger threatening mankind, and consider that the problem of preventing it takes priority over other international problems; I am in favor of disarmament and a strategic balance, I support the SALT II agreement as a necessary stage in disarmament negotiations. I am against any expansion, against Soviet intervention in Afghanistan, but in favor of aid to refugees and the starving throughout the world. I regard as very important an international agreement on refusal to be the first to use nuclear weapons, concluded on the basis of a strategic balance in the field of conventional weapons.

I do not make it my task to give special support to the viewpoint of Western governments, or anyone else, but express precisely my own viewpoint on matters causing me anxiety. As for the mildness of the measures taken against me, they are not as severe as the terms of imprisonment lasting many years for my friends and scientists—prisoners of conscience Sergei Kovalev, Yuri Orlov, Anatoly Shcharansky, Tatyana Velikanova, Victor Nekipelov—nor as the fate of those awaiting trial—Alexander Lavut, Leonard Ternovsky, Tatyana Osipova, and many others.

But my banishment, without trial in infringement of all constitutional guarantees, the isolation measures applied, interference of the KGB in my life, are completely illegal and inadmissible as an infringement of my personal rights and as a dangerous precedent of the actions of the authorities, who are casting aside even that pitiful imitation of legality in the persecution of dissidents that they displayed in recent years. Only a court has the right to establish that a law has been infringed and to define the manner of punishment. Any deliberations about culpability and mercy without a trial are inadmissible and against a person's rights.

Therefore, I insist on a public trial, and attach fundamental importance to this.

February 22, 1980

ANDREI SAKHAROV

The Human Rights Movement in the USSR and Eastern Europe: Its Goals, Significance, and Difficulties

The use of the word "movement" in the title of this article should not bring to mind any sort of organization or association, much less the concept of a party. We are simply speaking of people who are united by a somewhat common point of view and method of action. Being one of these people (a so-called "dissident"), I do not in any way aim to play the role of an ideologue or leader. Every one of my public statements, including this one, reflects my own personal opinion in regard to questions which concern me deeply.

The sociopolitical ideology which gives first priority to human rights is, in my opinion, the most reasonable in many respects, if considered within the framework of the relatively narrow set of problems which it places before itself. I am convinced that ideologies based on dogmas or metaphysical precepts, or those which rely too heavily on the contemporary makeup of their societies, cannot be responsive to the complexity, sudden changes, and unpredictability of human development. The imperative and dogmatic concepts of all types of world reformers, as well as the irrational mirages of nationalism and national

Written at the invitation of *Trialogue,* a publication of the Trilateral Commission. *Translated by Ludmilla Thorne.*

socialism, have so far been realized by violating the internal freedom of people and inflicting direct physical harm—embodied in the twentieth century in the horrors of genocide, revolutions, international and civil wars, anarchistic and state-inspired terror, and the hells of Kolyma[1] and Auschwitz.

Communist ideology, with its promise of creating a world society based on social harmony, labor, material progress, and future freedom, has, in fact, been transformed by governments which call themselves socialist into an ideology of a party-bureaucratic totalitarianism, leading in my view to the deepest historical dead end.

Moreover, at the present time there no longer exists a pragmatic capitalist philosophy of reasonable individualism, at least not in its pure form. The various upheavals of the twentieth century, such as the Great Depression, destructive wars, and the specter of an ecological and demographic catastrophe, have demonstrated its inadequacy.

I believe that technical-economic progress is a supremely positive factor in our social life which to a large measure lessens the problem of distributing material wealth. At the same time, however, I acutely feel the dangers which are tied to this kind of progress, and recognize the inadequacy of a technocratic ideology in solving life's many-faceted complex of problems.

In contrast to the imperative nature of the majority of political philosophies, the ideology of human rights is in essence pluralistic, allowing various possible forms of social organization and their coexistence. It also offers the individual a maximum freedom of choice. And I am convinced that precisely this kind of freedom, and not the pressure exerted by dogmas, authority, traditions, state power, or public opinion, can ensure a sound and just solution to those endlessly difficult and contradictory problems which unexpectedly appear in personal, social, cultural, and many other aspects of life. Only this kind of liberty can give people a direct sense of personal happiness, which after all comprises the primal meaning of human existence. I am likewise convinced that a worldwide defense of human rights is a necessary foundation for international trust and security; it is a factor

---

[1]Kolyma is a region in northeastern Siberia known for its complex of concentration camps. —Trans.

which can deter destructive military conflicts, even global thermonuclear conflicts which threaten the very existence of humanity.

In the postwar era, the ideology of human rights found its most
consistent expression in the UN Universal Declaration of Human
Rights, in the various human rights movements, and in Amnesty
International's global campaign for the amnesty of all prisoners of
conscience.

The ideology of defending human rights played a very special role
in the social movements in the USSR and in the countries of Eastern
Europe. This phenomenon is tied to the historical experiences of these
peoples, who within the life span of one generation lived through a
stormy, brief period of intoxication with communist maximalism (this
applies primarily to the USSR), with its accompanying intolerance,
dogmatism, general destruction, suffering, and the crimes committed
by both the Whites and the Reds in the name of that which they
considered to be their great goal. Then followed the bloody horror of
Stalin's fascism, which destroyed tens of millions of lives and which
slowly transformed itself into the present, stable phase of a party-state
totalitarianism. With this sort of experience behind us, it is natural
for us to embrace an ideology which first of all places emphasis on
the defense of concrete individuals and concrete rights, by methods
which are in principle nonviolent and nondestructive. It is an ideology which is based on the observance of laws and international documents which have been signed by our governments. The closeness
in ideas and even in methods used by dissidents to wage the human
rights struggle in the USSR and countries of Eastern Europe makes
it possible to speak of a unified human rights movement, in spite of
the absence of organizational ties between the movements in these
countries and the virtual inability to communicate—correspondence,
the use of the telephone, and mutual visits are completely blocked
by the authorities.

I would like to point out that one of the ways in which the authorities of these countries have reacted to such an absolutely lawful
and constructive position taken by the dissidents was to violate their
own laws—particularly during trial proceedings—to use with ever-
increasing frequency underground methods of provocation, and in some
instances even to commit acts of terror against certain individuals both

within their own countries and abroad. And these unlawful actions taken by the authorities have in turn strengthened the lawful orientation of the dissidents.

In Czechoslovakia the defense of human rights was a vital element in the "Prague Spring," and more recently, it formed the basis of the widely known Charter 77, which in its direction and spirit is very close to numerous documents of the human rights movement in the USSR and other countries of Eastern Europe.

In Poland, the Workers' Defense Committee came into being, as well as other associations. Ten years ago, a number of groups appeared in the USSR as a reaction to unjust trials and other violations of human rights. Among these were the Initiative Group for the Defense of Human Rights, the Human Rights Committee, and more recently, the Helsinki Watch Groups. The most important plateau in the formation of the human rights movement in the USSR was the creation of the remarkable samizdat information bulletin *A Chronicle of Current Events*, which, in spite of countless repressions and indescribable difficulties, has been coming out regularly for ten years now, with its traditional epigraph—the text of Article 19 of the Declaration of Human Rights.[2] I believe that this journal reflects best of all the very spirit of the movement—its objectivity and apolitical and pluralistic nature, its striving for accuracy and truthfulness, and its foremost interest in concrete violations of human rights and concrete fates of those who have become the victims of injustice.

The human rights movement in the Soviet Union and in the countries of Eastern Europe gives first priority to civil and political rights as a matter of principle, in contrast to the official state propaganda of these countries, which purposefully accentuates economic and social rights (thus even contradicting the founders of Marxist theory). I am convinced that under contemporary conditions it is precisely civil and political rights—the right to freedom of conscience and the dissemination of information, the right to choose one's country of residence and to live wherever one chooses within that country, freedom of religion, the right to strike, the right to form associations, and the

---

[2]"Everyone has a right to freedom of opinion and expression; this right includes freedom to hold opinions without interference and to seek, receive, and impart information and ideas through any media and regardless of frontiers."

absence of forced labor—which are the guarantees of individual liberty and give life to the social and economic rights of man, as well as to international trust and security. Civil and political rights are more systematically and more openly violated in totalitarian countries.

The key right, that of freely choosing one's country of residence, is violated with particular crudeness in the USSR and the German Democratic Republic with its Berlin Wall. The freedom to choose one's country of residence is important not only in family reunification (and I do not mean to diminish the importance of this), but it also gives the right in principle to leave a country which does not provide its citizens their national, economic, religious, political, civil, and social rights, and to return to it should there be a change in the personal life of an individual or in the general situation. This right must inevitably lead to general social progress.

In the USSR only an invitation from close relatives gives a citizen the right to apply for emigration. This limitation directly contradicts international law as described in the UN Covenant on Civil and Political Rights. Thus, in "one clean sweep" they write off a large number of people who wish to emigrate or to leave the Soviet Union even temporarily for economic, religious, national, political, cultural, medical, or other personal reasons. But even those would-be émigrés who have invitations—in particular, Germans, Jews, Lithuanians, Estonians, Latvians, Armenians, and Ukrainians—often encounter colossal difficulties. It is not by accident that there is such a word as "refusenik." It seems quite clear to me that the repeated instances of arrest and unjust convictions of those who strive to emigrate is an attempt to crush the emigration movement, to frighten and halt in midstream any potential émigré. The article on "treason" contained in the Criminal Code of the RSFSR and other republics includes, among the usual definitions of this crime, "escape abroad or refusal to return to the USSR." In accordance with this statute, hundreds of people have been sentenced to the most cruel punishments, and many have been placed in psychiatric prison hospitals.

For some time now, the fates of the convicted Jewish refuseniks— Shcharansky, Slepak, Ida Nudel, Goldshtein and Begun—have become widely known, as well as the cases of those who participated in the Leningrad hijacking affair.[3] There are particularly many denials to emigrate and all sorts of persecutions aimed at the Germans who wish to emigrate. (From the 1930s to the 1950s hundreds of thousands of Germans died during Stalin's deportations and repressions.) The fate of Peter Bergman, for example, whose peasant family has been trying to emigrate to Germany for three generations during the past fifty years, is most tragic.

In contrast to the generally accepted norm regarding freedom of movement within a country (Article 13 of the Universal Declaration of Human Rights and the corresponding article in the Covenant on Civil and Political Rights), in the USSR there is a passport system with an obligatory residency permit, the so-called *propiska,* which is issued by the Ministry of Internal Affairs. Freedom of movement for those who live on the kolkhozes (collective farms) is even more limited. The Kolkhoz Decree does not guarantee the right to leave the kolkhoz, which virtually reduces tens of millions of people to serfs. The fact that some of them nonetheless do, by one means or another, attain permission to leave the collective does not change the intolerability of the general situation.

A special series of human rights violations in the Soviet Union is connected to nationality problems. The Crimean Tatars, who along with many other nationalities became the victims of Stalin's genocide, are even now subject to the discriminatory prohibition of returning to their homeland in the Crimea. In 1944, when old men, women, and children (the younger men were at the front) were exiled from the Crimea, almost half of all the Crimean Tatars died. The humiliation and cruelty to which those families wishing to return to the Crimea are subjected is beyond description. Refusals to grant residency permits, imprisonment for violating residency rules (special permission must be received from the Ministry of Internal Affairs), denial of the necessary documents for buying houses, the destruction of homes

[3]In December 1970, eleven individuals, mostly refuseniks, were tried for treason, theft of state property, anti-Soviet agitation, and other charges, for the planning of a hijacking to Sweden. The key defendants, E. Kuznetsov and M. Dymshitz, were initially sentenced to death, but later received commuted sentences of fifteen years. The other defendants received sentences ranging from four to fifteen years.—Trans.

already bought (thus leaving entire families with children and old people on the street), forced evictions, inability to get work—all of these are part of a consistent discriminatory policy which Crimean Tatars must face.

In the summer of 1978 the Crimean Tatar Musa Mamut committed an act of self-immolation, in an effort to draw attention to the tragic situation of his people. As he was being taken to the hospital, dying, he said, "Somebody had to do it."

The acuteness of the nationality problems in the USSR is underlined by the severity of political repressions in the various national republics—in the Ukraine, in the Baltic areas, in Armenia, and elsewhere. Prison sentences in these republics tend to be particularly severe, and the reasons for imprisonment are even more flimsy than usual.

The Soviet Constitution formally proclaims freedom of conscience and the separation of church and state. But in reality, those churches which are officially recognized find themselves in the humiliating position of being totally dependent on the government in the administrative and material sense. They are deprived of the right to sermonize and to dispense charity, and their priests and elders are appointed by the Soviet organs.

Under these conditions, it is necessary to praise the clandestine nonconformity of the many rank-and-file clergy and believers of these churches. Those who oppose their church's dependency on the authorities are subject to particularly cruel persecution—the separation of children from their parents, commitment of believers to psychiatric hospitals, arrests, convictions, confiscations, and even terroristic acts— none of which are ever investigated.

Recently we were all shaken by the arrest of Vladimir Shelkov, the eighty-three-year-old leader of the Seventh Day Adventists, who had already spent more than twenty-five years in prison. Repressions which the followers of this church suffer for their religious activities are particularly ruthless. Very often these people are forced to live a marginal existence. The situation of the independent wing of the Baptist Church, the Uniates, the Pentecostalists, the so-called True Orthodox Church, and others is no easier.

In the Baltic republics and in the western parts of the Ukraine, religious persecution often assumes an antinationalist character. In Lithuania, for instance, the Catholic Church functions under great limitations, and its anonymous journal, *The Chronicle of the Lithuanian Catholic Church,* is hampered, its publishers and distributors persecuted. So far, I have been speaking about the religious situation in the USSR, which is particularly intolerant. It is well known that in some countries of Eastern Europe, the heroic efforts of believers and certain Church leaders—such as Cardinals Mindszenty in Hungary and Wyszynski in Poland—have contributed to the establishment of a much more normal situation. The authority which the Church enjoys in these countries is one factor that contributes to the lessening of totalitarian pressure on the individual.

Emigration for religious reasons constitutes a special problem. For several months now, two Pentecostal families, the Vashchenkos and Chmykhalovs, have voluntarily confined themselves to the American Embassy in Moscow. For over sixteen years they have tried to emigrate from the USSR, after having experienced all possible forms of persecution, including prison. Now, their own local newspapers accuse them of being "spies" for foreign powers, and who knows, perhaps the same fate awaits them as that of Shcharansky, should they decide to leave the territory of the embassy, which is guarded day and night by the KGB. Many others of their faith have been trying unsuccessfully to emigrate (in some cases entire Pentecostal parishes), as well as numerous Baptists and other believers.

Along with the right to freely choose one's country of residence, the image a society projects is determined most powerfully by the right to freedom of conscience and dissemination of information. These basic rights are contradicted by certain articles in the criminal code of the Soviet republics,[4] making it possible to persecute for just those nonviolent and lawful acts that occur in all democratic states. Hundreds

[4]Article 70 ("Anti-Soviet Agitation and Propaganda") and Article 190-1 ("Dissemination of Fabrications Known to Be False Which Defame the Soviet State and Social System") in the RSFSR Criminal Code.—Trans.

of prisoners of conscience are incarcerated on the strength of these articles. One of them is the prominent biologist, and my close friend, Sergei Kovalev, who was also one of the editors of A *Chronicle of Current Events*.

The political trials in the USSR and Eastern Europe under charges of this nature are conducted with the crudest violations of the accused's rights to examine the entire record of the preliminary investigation of his case, to protection against fabricated slander and closed trials. No one, with the exception of the closest relatives of the accused, is allowed into the courtrooms of nominally "open" trials, and at many of the recent trials, even the wives and mothers of the accused could not be present. Indeed, there is much to be concealed at these trial proceedings, as well as about what happens in the camps and prisons—but about this, I shall speak later.

Not long ago the attention of the entire world was focused on the lawless trials of members of the Helsinki Watch Groups—namely, the trials of Orlov, Ginzburg, Shcharansky, Petkus, Lukyanenko, Kostava, and before them, Rudenko, Tikhy, Marinovich, Matusevich, Gajauskas, and others—who were tried on the basis of these same articles.

In the last months the Moscow Helskinki Watch Group has issued a series of important documents. I joined in some of these, including the group's statement of October 30, 1978, which demands the nullification of Articles 70 and 190-1 of the RSFSR Criminal Code and that part of the article on treason (Article 64) which classifies as treasonable any attempt to leave the country.

The conditions in which one and a half million prisoners, including hundreds of political prisoners, are confined in Soviet camps and jails undoubtedly constitutes an inadmissible violation of human rights.[5] Prisoners suffer forced labor under severe conditions; and for the nonfulfillment of unattainable norms they are punished, most often by being placed in punishment cells where they are tortured by hunger and cold. There is no decent medical care, and they are exposed to constant provocations and harassment on the part of the administration. That is their life. During the press conference which was held

---

[5]The figures above are approximate; no exact statistics are available.

on October 30, 1978—which, since 1974, has been traditionally regarded as "the day of the political prisoner"—I transmitted to foreign correspondents a letter from the special-regime concentration camp in Sosnovka, in which these conditions were described with impressive exactitude and authenticity.

Among the rights which are extremely important for any normally functioning society, and which are not realized in the USSR and in the countries of Eastern Europe, are the rights to strike and to form associations independent from the authorities. Based on the example of these rights, it becomes increasingly clear that without the existence of political and civil rights, social and economic problems cannot be effectively solved. Soviet propaganda claims that our country is a fully developed socialist state with a maximum concern for the individual. But reality is far from these boisterous assertions: there is a tremendous social inequality between the working masses—especially such professionals as white-collar workers, doctors, and teachers—and the so-called bosses, who have countless privileges. This inequality is particularly painful in light of the extremely low standard of living, in spite of the fact that our country is relatively developed, in the economic sense. Allow me to cite a few figures. The average pay of a Soviet citizen is about 150 rubles per month, but some salaries are 80 and even 70 rubles, even in Moscow, where salaries are higher than in the provinces. The maximum pension is 120 rubles per month, but there are many different kinds of personal pensions, and the minimum comes to about 40 rubles. A single mother receives 5 rubles per month; but if the income of a given family comes to less than 50 rubles per person, the monthly allowance per child is 12 rubles, but this support is dispensed only up to age 8. In the majority of cities even the most essential items are lacking, in particular meat, medicines, and numerous basic manufactured products. People travel to Moscow from all corners of the country, spending their money, time, and effort just to obtain the basic necessities of life.

Humanity now faces a series of complex problems which threaten normal life and the happiness of future generations. These problems threaten the very existence of civilization. One of the more insidious dangers, and one difficult to prevent, is the spread of totalitarianism, which threatens humanity's progressive and free development. This is the very danger that the struggle for human rights directly counters. The ever-widening understanding of this fact has been reflected in recent years in such historical events as the signing of the Helsinki Final Act by thirty-five heads of state, where the inseparable link between international security and the observance of basic human rights was established. The same shift in public opinion was reflected in President Carter's articulation in January 1977 of a policy based in principle on the defense of human rights throughout the world as the moral basis of United States policy. In this conception, the global character is particularly important, that is, the attempt to apply the same legal and moral criteria for human rights violations to every country in the world—to Latin America, Africa, Asia, and the socialist countries, and to one's own country. I know of some important and fruitful results of this policy in South and Central America, and in other areas. I am not at all inclined to underestimate the importance of waging a human rights struggle everywhere such violations occur, not striving to limit this struggle only to the USSR and Eastern Europe. To eliminate the suffering taking place today is more important than anything else, and it is not at all important whether that suffering is near or far, in the geographic or national sense. But at the same time, I also wish to emphasize that the crux of the threat of totalitarianism lies in the USSR, and this too must be taken into consideration.

I believe that President Carter's principled position responds to the demands of our time and to the democratic traditions of the American people. It can further the unification of all democratic forces in the world, and it bears a historical significance which cannot be canceled by certain inaccuracies in the concrete execution of this policy. I consider it very important that the principled position put forth by the United States Administration regarding the defense of human rights receive even broader support, as well as those initiatives aimed at strengthening the position of the U.S. This is necessary for the U.S. to successfully carry out its leadership role in the Western world, to counterbalance the offensive policy of totalitarianism.

I have in mind here even such strictly domestic issues as the energy program and the fight against inflation. It seems to me that the discussion of key problems in the currently tense situation should be conducted above party politics and above other internal differences. Also in need of our support are such key issues of international life as the peaceful settlement between Egypt and Israel, which is in the interest of all the peoples of the Middle East and the world as a whole. A problem which is more modest in appearance, but important for the economic and political independence of the West, is the effort toward the peaceful uses of nuclear energy. I was recently distressed to learn about the negative results of the referendum held in Austria on this question.[6]

The American people are freedom-loving, generous, active, and energetic (at least, that is the image that I have of them), and they will undoubtedly be able to overcome those problems which now face them and the entire world. One especially significant reflection of the evolution in the social climate of the world has been the series of political amnesties in many countries—and many of these countries are far from democratic. There was an amnesty in Yugoslavia, Indonesia, Chile, and Poland. An amnesty is also scheduled to take place in Iran and in the Philippines, as well as in several countries of Latin America. The human rights struggle in the Soviet Union and the countries of Eastern Europe was one of the factors which contributed to these events—the freeing of thousands of people.[7]

At the present time, that small handful of dissidents whom I know personally are going through a difficult period. Many wonderful, courageous people have been arrested. The campaign of slander and provocation is intensifying, which in part is directly orchestrated by the KGB, but to some degree is used by, or reflects the division, discontent, and disillusionment among, some of the human rights activists and the circles close to them. Life is complex—and under these circumstances personal resentments and ambition have driven some people to actions and statements of a questionable nature. The number of active participants in the human rights movement in Moscow and in the provinces has probably diminished somewhat.

[6]On November 5, 1978, Austrian voters rejected a plan to open the country's first nuclear power plant at Zwentendorf.
[7]In most cases, these amnesties have been partial ones.

And yet, I believe that there is no basis for stating that the human rights movement has been defeated. It is one of those areas where arithmetic has very little relation to the thing at hand. During the past few years the human rights struggle in the USSR and Eastern Europe has substantively changed the moral and political climate of the entire world. The world has not only received a wealth of information, but has also believed in it. And this is a fact which none of the repressions or provocations of the KGB can change. It is the historical reward earned by the human rights movement. Now as before, the sole ammunition of this movement is publicity-free, accurate, and objective information. This weapon continues to be effective. It is also quite evident that as long as the conditions and the goals of the human rights struggle remain, new people, motivated by certain circumstances in their lives and their own spiritual drive, will take the place of those who have left. And this too cannot be prevented by the repressions of those who are in power. On the contrary, a cessation of repressions would result in a major improvement in the authorities' position.

What do I expect from people in the West who sympathize with the human rights struggle? It is true beyond a doubt that their help is very important. And in this connection I should like to focus my attention on a few questions which are now being debated. The great deal of attention directed toward human rights problems in the USSR and the countries of Eastern Europe, especially following the period of trials in the spring and summer of 1978, is an extremely important factor on which I place much hope. But expanded possibilities also demand extreme accuracy and judiciousness in action, keeping all the possible consequences fully in mind.

In the Western press the thought has sometimes been expressed that the strategic arms limitation talks, in whose success the Soviet Union is interested, as is the entire world, have opened up possibilities of applying pressure on the USSR on the question of human rights. In my opinion, such a viewpoint is not correct. I believe that the problem of lessening the danger of annihilating humanity in a nuclear war carries an absolute priority over all other considerations. I believe that the principle of practicably separating the question of disarmament from other problems, as formulated by the United States

Administration, is completely correct. Consequently, the strategic arms limitation talks must be considered separately; and considered separately, we must ask ourselves whether it will lessen the danger and destructive power of a nuclear war, strengthen international stability, or prevent a one-sided advantage for the USSR or a consolidation of its already existing advantages. Such a separate, practical approach does not negate, of course, the undoubted fact that a durable international security and international trust are impossible without the observance of the basic rights of man, specifically political and civil rights. It should also be pointed out that the West should not consider the cutting of military expenditures as the main goal of arms limitation. The basic goals can only be international stability and the elimination of the possibility of a nuclear war.

Another problem widely discussed in the Western press concerns the use of boycotts—scientific, cultural, economic, and so forth—as a means of applying pressure on the USSR for the purpose of freeing at least some political prisoners. After the trials of Orlov, Shcharansky, and Ginzburg, many Western scientists refused to take part in the scientific seminars and conferences held in the USSR. Some scientific associations refused altogether to cooperate with Soviet scientific institutions. I welcome such boycotts as a means of expressing the protest of world public opinion against the violations of human rights in the USSR. The same applies to economic boycotts, for example, the refusal to sell computer technology or oil-drilling equipment. The Soviet Union and other totalitarian countries must know that the politics of defending human rights is not simply a beautiful phrase used by Western politicians, but an expression of the people's will in Western countries, and that the continuation of human rights violations is irreconcilable with the continuation and expansion of détente. The same thought can be suggested by Western businessmen, politicians, athletes, lawyers, and many others who have dealings with the leaders of totalitarian countries.

However, the problem of boycotts is complex and contradictory. No doubt the question of prestige in the world political arena, the struggle to attain and keep power, especially in the context of behind-the-scenes struggles, and the very traditions of a strong power do not allow the leaders of totalitarian states to react directly to pressure

exerted against them. It is also certain that at the same time, boycotts
weaken realistically useful contacts, and diminish the number of le-
vers which can be used to apply pressure in the future. Therefore, in
such complex matters it is impossible to give one single answer which
could be applied in all cases. I can only express a few general considerations. It seems to me that with a few rare exceptions, it is best to
avoid boycotts with ultimatums. That is, it should not be indicated in
an obvious manner that the boycott will cease only if the totalitarian
regime undertakes certain concrete steps. In such a case, a boycott
will demonstrate the one side's interest in a particular cause, and at
the same time create a situation where the opposite side is pushed
into a "dead end" from which it cannot extricate itself without losing
face.

I am also convinced of the necessity of combining various and
impressive public campaigns with an energetic and thoughtful quiet
diplomacy. The exchange of political prisoners can be an important
arena of action for quiet diplomacy. I have already written that I do
not understand and do not accept the contentions against such exchanges which have been expressed in the West. It seems to me that
in some cases, this is practically the only realistic way to tear people
out of the hell of the camps and prisons. Even if this method can help
only very few people, still, it is a breakthrough, and it assuredly does
not harm those who remain behind. Prisoner exchanges also do not
undermine the authority of human rights organizations, such as Amnesty International, whose goal is to achieve a worldwide political
amnesty.

The proper attitude toward the forthcoming Moscow Olympics is
a separate problem. My position coincides with the opinion expressed
in the letter of the Moscow Helsinki Watch Group, which I also signed.
It was sent to the International Olympic Committee and to its president, Lord Killanin. The authors of this letter note the human rights
violations which exist in the USSR and warn that the Soviet authorities, in complete disdain of Olympic principles, intend to limit the
contacts between people at the upcoming Olympic games. The authors urge that this not be permitted and demand an end to persecution for nonviolent actions in support of human rights, i.e., for religious activity and the attempt to act on the right to freely choose

one's country of residence and place of domicile within that country. They call for the liberation of all prisoners of conscience. The authors of the letter state that they attach great significance to the forthcoming Olympic games, and ask that their letter be brought to the attention of the national Olympic committees and sports organizations in various countries, in order that every participant in the forthcoming Olympics may express his concern regarding the questions raised. Unfortunately, up to now we have not been informed of the Olympic Committee's reaction to this document.

The ideology of human rights is probably the only one which can be combined with such diverse ideologies as communism, social democracy, religion, technocracy, and those ideologies which may be described as national and indigenous. It can also serve as a foothold for those who do not wish to be aligned with theoretical intricacies and dogmas, and who have tired of the abundance of ideologies, none of which have brought mankind simple human happiness. The defense of human rights is a clear path toward the unification of people in our turbulent world, and a path towards the relief of suffering.

Moscow, November 8, 1978
(Published in *Trialogue* magazine, January 1979)

ELENA BONNER AND ANDREI SAKHAROV

Two Appeals

Tolyá [Anatoly] Marchenko has been arrested again. This news is so frightful that it is hard to absorb it into one's consciousness. Marchenko's life is familiar to readers of his splendid books, *My Testimony* and *From Tarusa to Siberia*. They are a burning indictment of the stupid cruelty of the repressive machine, and at the same time a testimonial to the true greatness of the human spirit, to the pride and honesty of a living, suffering man who opposes that machine. A blue-collar worker and writer who has told the truth (so important for all of us) about today's Soviet penal camps, he is one of those of whom the country and people who gave him to the world can rightly be proud. Today, when the vengefulness of his jailkeepers has again descended upon him, our hearts go out to him and his family. We ask all honest people of our country and the world to do everything in their power to defend and help him.

Gorky, March 22, 1981   Elena Bonner
                                                          Andrei Sakharov

The sentencing of Tanya [Tatyana] Osipova, an infinitely honest, courageous young woman who selflessly took upon herself a concern for innocent sufferers, for justice and the public disclosure of information, is yet another example of the monstrous cruelty of the repressive organs, another lawless act for all the world to see. Honest people, do everything possible for her release! Appeal to chiefs of the states signatory to the Helsinki Act, to Amnesty International, to scientists, writers, workers.

ANDREI SAKHAROV

Gorky, April 3, 1981

ANDREI SAKHAROV

How to Preserve World Peace

In spite of all the differences in their histories, the overwhelming majority of people in all countries want peace. The people of the Soviet Union and the United States are no exception. War means suffering and death, the loss of loved ones, cruelty, separation, poverty, destruction, hunger, and sickness.

This is how war has always been, and that is how hundreds of millions of people who are still mourning their dead saw it in the two world wars and in the many "small" but no less cruel wars of this century. A third world war could be even more horrible. The equivalent of 13 billion tons of TNT concentrated into 40,000–50,000 thermonuclear and nuclear charges threatens the very existence of mankind, or at least of civilization.

I believe that the leaders of all of today's states cannot ignore the passionate will to peace common to all mankind. Gone are the days of medieval barons, who considered war proof of their chivalry and valor while, time after time, their peasants humbly resowed their trampled fields and rebuilt their burned-down huts. Today, both Reagan and Brezhnev—as individuals, alone with themselves—undoubtedly want peace for their peoples, their loved ones, for all the people on earth. This I sincerely believe.

But life is extraordinarily contradictory and complex. To our sorrow and our horror, there are very serious factors whose interactions are not controllable and which push the leaders of some states into dangerous actions that bring the entire world to the brink of catastrophe. The false logic required to retain power is one factor which hinders the necessary compromises and reforms. (I am speaking here of the power of the Communist Party in the Soviet Union, but there are other examples as well.)

Another risk factor is expansion, the struggle for increased spheres of influence. Such struggles sometimes take place because of an erroneous understanding of the problem of security. This is true both for the USSR and the U.S.A., but more often than not expansion occurs because of the false and dangerous messianism of the USSR. Direct armed interference in the affairs of the countries in one's camp (the USSR's), if those countries have chosen the path of reform, is a most dangerous action, one which destroys the most important foundations of international relations and international equilibrium. The fear and distrust intensified by the closed nature of the socialist world is another factor here, as is the arms race, which has created the cancer of a military-industrial complex in both the USSR and the U.S.A. There is also the danger that small, local conflicts will escalate into large, global ones.

All of these factors are operating in an unprecedented context of global confrontation. Moreover, the socialist world and the West both possess peculiarities that increase the danger.

The USSR was born under the banner of world communism but, to a significant degree, has lost its ideological fire. Now the basic mood is the passivity and indifference of a people induced to drink, burdened and tired by constant economic difficulties. At the same time, they are completely loyal to the Soviet way of life, for all its lack of freedom, and to the rule of the Communist Party, which is seen as a mainstay of stability and peace. (To a significant degree, this is precisely why the sense of a supposed danger from the West is kept artificially heated up.)

Having lost the long view (the short-term view now means building private summer homes), the Party authorities continue traditional Russian geopolitics but now on a worldwide basis—taking advantage of the enormous resources of a totalitarian system, a unified and slanted but clever and consistent propaganda both inside the country and out. It quietly penetrates all the cracks in the West and employs subversive activities there; it exploits the increased though one-sided resources of its economy for unrestrained militarization.

One need not take seriously the figures on military expenditures cited by Soviet propaganda, for they are always lowered. In addition, the peculiarities of the Soviet economic system allow for a free and uncontrolled redistribution of resources and seem to create armaments without paying anything for them (but meanwhile there is no cotton or meat in the stores; however, it is an illusion to think that this to any degree impedes Soviet military development or causes any open disturbances of any significance among the populace, especially in a country which has been periodically stricken by famine throughout its history but which now has had none for several decades).

One aim of Soviet foreign policy is the disorganization and intimidation of the West, the exploitation of the West's technological and economic resources while threatening it with rockets. All this playing with fire, the height of which was the tragic error of invading Afghanistan (where the majority is now undergoing forcible conversion to the socialist faith), may have been due to the absence of social control over the actions of our leaders and the almost totally closed nature of our society.

Foreign radio broadcasts are being jammed again. *Glavlit* (the censorship) has a 100-page list of forbidden topics, from statistics on alcohol consumption to any mention of Stalin's crimes. The number of tourists from the USSR going abroad is less than that of little Denmark, and, like our tourists abroad, foreigners here are deprived of the opportunity of speaking freely with people. Again repressions are being stepped up against those who defend the rights of freedom of information and freedom of movement. The names of Tatyana Velikanova, Yuri Orlov, Sergei Kovalev, and Anatoly Shcharansky have become symbols of those repressions. And what will happen tomorrow?

At the same time, the West is utterly divided and pluralistic— which constitutes both its strength and its weakness in opposing totalitarian expansion. How easily Soviet propaganda launches mass, one-sided campaigns against the placement of American rockets (only!)

in Europe—and this at a time when an obvious disturbance of military equilibrium has occurred in that part of the world and that includes the balance of nuclear rockets.

Frequently, Western intellectuals, when speaking out against the arms race—itself a necessary thing to do—take a one-sided position which fails to take the realities into account, an attitude compounded by the anti-American feeling of many Europeans. How often business (for the sake of quick profits) and the press (for the sake of sensationalism, cheap popularity, circulation) oppose their governments' intelligent economic (e.g., energy) and foreign policies. At times one gets the impression that if the West were really like that, it could be taken by the totalitarian strategists with their bare hands. Actually that is not the case, but why create the temptation?

The continuous increase of nuclear threat in Europe from Soviet modernized medium-range missiles, and, in response to this, the stationing of new ballistic and cruise missiles planned by NATO, have become one of the major military-political problems of recent years. Undoubtedly this problem can only be solved provided that the dictates and demagoguery of the USSR are eliminated, and that the West is unified and ready to display the necessary steadfastness while remaining at the same time ready to compromise.

What should the U.S.A. and the USSR—government, people, press—do to preserve peace? To be quite brief, they must realize the risk factors in their mutual relationship, explain them to the people (here scientists and a responsible press can play a large role), and, by mutual efforts, attempt to eliminate those factors.

The government of the USSR should realize that any attempts to alter the equilibrium existing in the world, no matter what considerations are used as a cover, are inadmissible. Furthermore, there must be withdrawals wherever "a little extra has been grabbed." Soviet troops must be withdrawn from Afghanistan, and elections under the control of UN forces must be held.

Actions against the Polish people and their lawful aspirations for economic and social justice, for pluralism and democratization within the framework of the existing system in Poland, would be even more dangerous, tragic, and destructive to world equilibrium. I hope that the Party authorities of the USSR, understanding the consequences, will refrain from irreparable actions. In this question, as in that of Afghanistan, the firm and unambiguous position of the West, its governments and its public opinion, is very important.

The government of the U.S.A., the government of the USSR, and all the members of the UN ought to undertake a wide-range joint program to peaceably combat the economic and social problems of the Third World countries, taking into account their specific character and their national traditions; and this should be done for the sake of peace on earth and not for influence, profit, or cheap raw materials. Such altruism is rarely encountered in the world, but today we cannot do without it.

The resolution of all international differences by means of negotiations is a necessary condition for peace. The USSR and the U.S.A. have repeatedly proclaimed their fidelity to this principle. I consider it especially important that public opinion be fully informed and know the points of view of both sides on the critical problems on which the fate of the world depends. Perhaps special international bulletins could be published in which both sides would state the information they considered important and their evaluation of it, with a government guarantee of its distribution (the latter being especially important in a closed society like ours).

This proposal, one of many possible, is only an example of an exceedingly serious common problem. The most important conditions for international trust and security are the openness of society, the observation of the civil and political rights of man—freedom of information, freedom of religion, freedom to choose one's country of residence (that is, to emigrate and return freely), freedom to travel abroad, and freedom to choose one's residence within a country.

The human rights proclaimed by the Universal Declaration in 1948, and again confirmed by the Helsinki Accord in 1975 as being part and parcel of international security, are rights which continue to be flagrantly violated in the USSR and in other countries, particularly those of Eastern Europe. It is necessary to defend the victims of political repression (within a country and internationally, using diplomatic means and energetic public pressure, including boycotts). It is also necessary to support the demand for amnesty for all prisoners of conscience, all those who have spoken out for openness and justice without using violence. The abolition of the death penalty and the unconditional banning of torture and the use of psychiatry for political purposes are also necessary. Other necessities include the demand for a series of legislative and administrative measures, including the abolition of censorship, the facilitation of emigration, travel, and so on. I turn for support for these demands to my colleagues in science, Soviet and Western; to the public figures and statesmen of all countries; to the people of the entire world. The governments and the public of all countries must insist on the unconditional and complete fulfillment of the humanitarian obligations the USSR has taken upon itself, in particular, in the UN's International Conventions on Human Rights and in the Helsinki Accords. This is a condition for being able to trust the signature of the USSR.

Parallel with these political, economic and legal conditions for the preservation of peace, the ceasing of the arms race and disarmament are of decisive importance. I am convinced that it is again necessary to return to the SALT II treaty, which is, in my opinion, a definite step forward and, more important, a necessary stage in facilitating further talks in this vitally important direction (some revisions may be required). At the same time, there is no doubt that deep-rooted progress in averting the nuclear menace is possible only in conjunction with the maintenance and, if necessary, the restoration of the balance of power between the West and the socialist camp in the area of conventional weapons—provided that public opinion in the West realizes the seriousness of the totalitarian threat and achieves greater psychological readiness to meet it.

In my opinion, new treaties (and an inspection system to supplement them) are needed to ban the use, production, and development of all chemical, incendiary, and bacteriological weapons.

Efforts also are needed to limit and reduce conventional weapons. In particular, it is necessary to limit the shipment of arms from the industrially developed countries to those areas of the world where armed conflicts are occurring or about to occur. Unfortunately, such shipments have, in the course of recent decades, taken place on a large scale and have created a danger to peace. Perhaps this is the seed from which a third world war will grow. This is a very complex problem, one interwoven with the necessity of resisting aggression and expansion—totalitarian expansion in particular—and with questions about the need of aiding allies and friends. A problem of this sort can be resolved only on a bilateral basis. I believe that with a sophisticated political approach and the goodwill of all sides, a solution can be found.

Again I turn to the U.S.A. and, in particular, to the USSR, since the resources of a totalitarian system in this field are especially great. There must be a decisive renunciation of all forms of subversive activity, including the use of an influenced press; the direct and indirect bribing of members of the press, business, and politics; and, especially, there must be a renunciation of support for international terrorism, which is a criminal and destructive weapon and one which is double-edged.

Terrorism has its roots in the past, and its motives are highly diverse. But no goals, national or social—not even vengeance for horrible crimes already committed—can justify the cruel murder of innocent people, including children, the taking of hostages, torture, or blackmail. Terrorism means cruelty and crime, and should inspire only aversion.

I hope that good sense will prevail and this contemporary nightmare will cease to be a menace to humanity. The use of terrorism by governments, whether directly or indirectly through intermediaries, or the support of it, is inadmissible. The most serious international decisions may be required to halt irresponsible acts by certain governments and political figures.

In rejecting hostage-taking by terrorists, one must not make a hostage out of the need for preserving peace for the future of mankind. Here I am in agreement with the statement from the UN Commission's report. Perhaps mutual nuclear terror is still keeping the world from World War III, but this distorted and wasteful balance of fear is becoming increasingly unstable. Political errors, new technological achievements by one side or the other, and the spread of nuclear weapons threaten to upset that balance at any moment. We must achieve a balance of power without the factor of nuclear terror by limiting ourselves to conventional weapons—whatever that might cost in economic and social terms—and public opinion must be mobilized in support of those efforts.

The halting of expansion, the regulation of conflicts through negotiations, the creation of an atmosphere of trust and openness, the maintenance and restoration of a balance of conventional weapons— only under these conditions can there be progress in reducing conventional and nuclear weapons and in reducing the danger of war. Under those conditions, it will be possible to take the exceedingly important step toward removing the threat of thermonuclear annihilation from mankind. That step would be taken by concluding a treaty against any first use of nuclear weapons. In the long run, all this should lead to the complete banning of nuclear weapons. This is what we must strive for.

A quarter of a century ago, the explosions above the Pacific Ocean and the Kazakhstan Steppe marked mankind's entrance into a paradoxical epoch of mutual thermonuclear terror. But only equilibrium based on reason—not on fear—is the true guarantee of the future.

Published in *Parade* magazine, August 16, 1981

ANDREI SAKHAROV

Three Letters Concerning the Hunger Strike

The following statement by Andrei Sakharov was taped in Gorky on November 15, 1981, and played by his wife, Elena Bonner, for correspondents in Moscow on November 16. It has been translated by Khronika Press from a transcript supplied by Efrem Yankelevich.

On November 22, my wife and I will begin a hunger strike for the reunion of our son Alexei and our daughter-in-law Liza.

Three and a half years ago Alexei, at my urging, was persuaded to emigrate. I assumed that Liza would follow shortly after. Alexei departed alone. Liza and Alexei have been separated ever since, and Liza has been subjected to persecution, libel, threats, and blackmail.

This past May she was officially refused permission to leave the USSR on the baseless pretext that there was insufficient justification for reunion. The authorities then made the illegal demand that she give a written promise to end her attempts to join Alexei. Liza flatly refused. A few days later, she was summoned to the KGB for an interrogation, which was conducted in a crude and threatening manner.

In 1981, Liza and Alexei were married by proxy according to the laws of Montana. Liza, six months after her last refusal, submitted a new application for reunion, accompanied by proof of her marriage. But in view of the fact that the Soviet consulate in the United States had refused to attest to invitations from her husband, she asked permission to leave for Israel to Alexei's close relative, as she had done in her first application.

My wife and I have decided on the extreme step of a hunger strike because of the responsibility we feel for the fate of two people dear to us and because we have exhausted all other means of helping them. The KGB, by making Liza a hostage for my public activity, has not only caused her and Alexei many years of suffering, but has changed this from a personal matter into a public, even a political, affair. The fight for Liza and Alexei has become a necessary, logical component of my long-term defense of human rights, of an open society, of law, of humaneness, of international security and confidence. It is important for me that those friends who are speaking out in my defense around the world understand my feelings on this subject. I hope that Liza and Alexei will also understand and accept our decision calmly and with a clear conscience.

The beginning of our hunger strike coincides with Brezhnev's visit to the Federal Republic of Germany, which inevitably lends it a political coloration. The KGB, not I, is responsible for this. We can oppose illegality, brutality, and cynicism only with firmness in our non-violent, open struggle. We count on the support of world public opinion and of our friends here and abroad, and on help from my scientific colleagues and from public figures and statesmen everywhere.

I must mention as a postscript that K., one of the two residents of Gorky permitted to visit me, was summoned to the KGB on November 11. Two topics were discussed: the theft of my car and our hunger strike.

On the subject of the car, two days after my telegram to Brezhnev and Alexandrov about my hunger strike, a KGB agent said, as if in passing, that my wife is getting ready to buy a new car. She simply abandoned the old one somewhere. It will probably be found in the spring.

The conversation on the second subject, the hunger strike, was more serious. The KGB agent told K. that my wife is trying to kill me off by the ingenious and dramatic means of a hunger strike. K. was forbidden to visit me during the hunger strike. He was told: "You have no business there." These statements by the KGB represent a direct threat. The KGB agent knew the conversation would be reported to me. That was the purpose of the conversation. And the message conveyed was: If Sakharov dies because of the hunger strike or through the intervention of the KGB, Elena Bonner will be held responsible. My death or murder will take place without witnesses. The world will be informed of the event by the KGB and will be fed a version favorable to the KGB. Those who understand the KGB's language will recognize the threat of murder.

I am making K.'s conversation with the KGB public in order to reveal the hidden causes and possible KGB links of any unexpected and dangerous event connected with our hunger strike.

Why not let Liza go instead of all these maneuvers and threats? Then no hunger strike would be needed.

The following letter was a reply to telegrams sent by Joel Lebowitz, past president of the New York Academy of Sciences, and Jeremy Stone, director of the Federation of American Scientists, asking Sakharov to terminate his hunger strike and assuring him that efforts on behalf of Liza would continue.

Dear Joel Lebowitz and Jeremy Stone,

Deep thanks for your energetic efforts, concern and attention. For more than two years I have been striving for a solution to a purely personal problem, a problem which is morally and legally beyond dispute. I have appealed to the head of our government, to the head of the USSR Academy of Sciences, to Soviet scientists, and to foreign colleagues and government figures. Now the sole possibility of ending our hunger strike is the departure of Liza, the discontinuance of government hostage-taking, dangerous as a precedent, and irresponsible, cruel and illegal. I can no longer believe in any kind of promises from the authorities not backed up by action! I ask you to understand and take this into account.

With esteem and thanks,

Sincerely,
ANDREI SAKHAROV

Gorky, November 27, 1981

We are deeply grateful to everyone who supported us in these hard times—to the statesmen, to the religious leaders and public personalities, to the scientists and journalists, to our dear ones and friends, to those whom we know and to those whom we do not know. There were so many—it is impossible to name them all.

It was a struggle not only for the life and happiness of our children, not only for my honor and dignity, but also for the right of every human being to be free and happy, for the right to live in accordance with one's ideals and beliefs, and in the final count it was a struggle for all prisoners of conscience.

Today we are happy that we did not cast gloom over Christmas and the New Year for our dear ones and for our friends throughout the world.

Wishing Liza a happy journey, I hope for the reunion of all who are separated, and I recall the wonderful words of Mihajlo Mihajlov that motherland is neither a geographical nor a national concept; motherland is freedom.

Sakharov's Protests on Behalf of Human Rights

In the case of A. D. Sakharov, the struggle to achieve peace and the struggle for the observance of human rights have always had a constructive character. He has not simply insisted, publicized, and explained what is threatening peace, but has proposed specific steps for eliminating dangers. He has not merely interpreted the essence and significance of human rights, but has expended all possible (and impossible) efforts to save specific individuals—acquaintances and strangers—from persecution.

We have almost completely omitted these matters, not because they seemed less important than general problems but simply because we feared that their enumeration would require many hundreds of pages. Unjust arrests, harsh sentences, persecution for desiring to emigrate from the country, deprivation of employment and residence permits, confinement in psychiatric hospitals, national and religious persecutions, discrimination in admission to colleges—all these things, embodied in the fate of individuals, demanded and inevitably found Sakharov's participation and support in the struggle against them.

He protested against Aleksandr Solzhenitsyn's expulsion from the country; he protested against Pyotr Grigorenko's being deprived of his citizenship; and at the same time he tried to get permission for Efim Davidovich to emigrate. He struggled for the right of the Crimean Tatars to live in the Crimea, and defended freedom of emigration for inhabitants of the Crimea, Moscow, Armenia, the Ukraine, the Baltic states, and Kazakhstan.

Sakharov became a man to whom people were drawn from all parts of the country in search of protection as to the last supreme court of honor and justice. This traditional Russian-Soviet phenomenon of "petitioners" absorbed almost all of his time and almost all of his strength. He always tried to help the people who came to him with their countless griefs, pains, and injuries. He could not do otherwise.

He was concerned about the plight of refugees from South Vietnam, and about the delivery of foodstuffs to Cambodia. He wrote the President of the United States about Angela Davis, called for Pablo Neruda's help, spoke out for Mihajlo Mihajlov, and defended the Polish workers and the Czechoslovak dissidents. But that Andrei Dmitriyevich should be primarily concerned with the persecution of people in our country was only natural.

Here is a partial list of the repressed persons for whose release A. D. Sakharov has struggled.

V. Abankin, V. Abramkin, P. Airikyan, G. Altunyan, A. Altman, A. Amalrik, Z. Antonyuk, E. Arutyunyan, K. Babitsky, V. Balakhonov, V. Bakhmin, A. Bergman, O. Berdnik, A. Berdnichuk, L. Bogoraz, A. Bolonkin, N. Bondar, V. Borisov, V. Brailovsky, N. Budulak-Sharygin, V. Bukovsky, G. Butman, B. Vail, T. Velikanova, S. Verkhov, G. Vins, P. Vins, O. Vorobev, Yu. Vudka, Yu. Galanskov, R. Galetsky, Z. Gamsakhurdia, B. Gajauskas, I. Gel, V. Gershuni, A. Ginzburg, S. Gluzman, P. Goldshtein, N. Gorbanevskaya, M. Gorbal, S. Gorbachev, N. Goretoi, P. Gorodetsky, I. Grivnina, P. Grigorenko, G. Davydov, B. Dandaron, Yu. Daniel, V. Delone, M. Dzhemilev, R. Dzhemilev, V. Dremlyuga, I. Dyadkin, A. Esenin-Volpin, R. Zagrobyan, V. Zalmanson, I. Zalmanson, M. Zand, I. Zisels, A. Zdorovoi, V. Zosimov, V. Igrunov, V. Jaugelis, R. Kadyev, I. Kalinets, V. Kalinichenko, I. Kalinets-Stasiv, I. Kandyba, S. Karavansky, A. Karapetyan, M. Kiirend, S. Kovalev, B. Kovgar, V. Kolomin, M. Kostava, A. Krasnov-

Levitin, E. Kuznetsov, Ju. Kukk, M. Kukobaka, A. Lavut, M.
Landa, V. Lashkova, V. Lisovoi, P. Litvinov, L. Lukyanenko,
A. Lupinos, K. Lyubarsky, L. Lyubarsky, M. Makarenko, V.
Malkin, S. Malchevsky, M. Marinovich, R. Markosyan, A.
Marchenko, N. Matusevich, O. Matusevich, Ma-Khun, Zh.
Medvedev, V. Meilanov, I. Mendelevich, I. Meshener, O.
Meshko, Mirauskas, D. Mikheyev, V. Moroz, A. Murzhenko,
K. Myatik, A. Navasardyan, A. Nazarov, R. Nazaryan, M.
Nashpits, V. Nekipelov, M. Niklus, I. Nudel, A. Ogorodnikov,
I. Ogurtsov, Yu. Orlov, M. Osadchy, V. Osipov, T. Osipova,
V. Pavlenkov, V. Pailodze, P. Paulaitis, E. Petrov, B. Per-
chatkin, R. Pimenov, M. Plakhotnyuk, P. Plumpa, L. Plyushch,
A. Podrabinek, K. Podrabinek, E. Prishlyak, E. Pronyuk, Putse,
B. Penson, V. Petkus, G. Rode, L. Roitburt, V. Romanyuk,
M. Rudenko, P. Rumachik, Saarts, N. Sadunaite, S. Sapelyak,
A. Safronov, I. Svetlichny, N. Svetlichnaya, E. Sverstyuk, M.
Semenova, I. Senik, A. Sergienko, F. Serebrov, L. Simutis,
A. Sinyavsky, Sinkiv, V. Slepak, S. Soldatov, N. Strokata, E.
Stroyeva, V. Stus, A. Tverdokhlebov, I. Terelya, A. Terleckas,
L. Ternovsky, O. Tikhy, Tovmasyan, A. Turik, L. Ubozhko,
V. Fainberg, V. Fedorenko, Yu. Fedorov, N. Fedoseyev, A.
Feldman, A. Khailo, V. Khaustov, M. Kheifets, A. Khnokh,
A. Chekalin, A. Chinnov, S. Shabatura, B. Shakirov, B.
Shakhverdyan, V. Shelkov, Yu. Shikhanovich, N. Shkolnik, I.
Shovkovy, M. Shtern, D. Shumuk, Yu. Shukhevich, A.
Shcharansky, V. Tsitlenok, A. Yuskevich, G. Yakunin.

Andrei Dmitriyevich Sakharov has sent letters to various Soviet agencies, has met with highly placed officials, has sent appeals to Western scientists, politicians, and international organizations, has sat in courtrooms, and, after 1971, has stood outside
buildings in which trials were being held, going for that purpose
to distant parts of our country. He has traveled to places of exile
and to the Mordovian camps, he has gone on hunger strikes, he
has held press conferences, he has granted interviews, and he has
written, written, written: letters, appeals, articles, giving all his strength

to saving people. In this hard struggle there have been few victories and many defeats. But he has not retreated, and he has not given up.

Andrei Dmitriyevich Sakharov's life work has become the courageous defense of every persecuted man and woman.

Biographical Notes on Contributors

Boris Altshuler is a scientist who met regularly with Sakharov at seminars and who spoke out vigorously in his defense after Sakharov's exile. A human rights activist, he lives in Moscow. Recently, he was fired from his job, and has been threatened with prosecution for his ties to Sakharov.

Larisa Bogoraz (1929–) is a linguist and human rights activist. She took part in the 1968 Red Square demonstration against the Soviet invasion of Czechoslovakia. She lives in Moscow.

Valery Chalidze (1938–), physicist and human rights activist who helped Sakharov found the Moscow Human Rights Committee in 1972, lives in New York and published the journal *The Chronicle of Human Rights in the USSR*. His books *To Defend These Rights* (1975) and *Criminal Russia* (1977) have been published by Random House.

Lidia Chukovskaya (1907–) is a writer and editor whose books include *Going Under, The Deserted House,* and a book about Anna Akhmatova; she has also written letters, articles, and poems. Expelled from the Soviet Writers Union in 1974 for her human rights activities, she lives in Moscow.

Evgeny Gnedin (1898–), a historian, was born into the family of one of the leaders of the left wing of the German Social Democratic

Party, grew up in Russia, worked in the People's Commissariat of
Foreign Affairs, and was arrested in 1939. He spent sixteen years in
the Stalinist camps. In 1980 he resigned from the Communist Party
in protest against Sakharov's exile to Gorky.

Philip Handler (1917–81), scientist and educator, taught for many
years at George Washington University. Apart from his many contributions to science, Handler served as president of the American National Academy of Sciences (1969–81). In 1973, he interceded with
the Soviet Academy of Sciences on behalf of Andrei Sakharov.

Sofia Kalistratova (1905–) is now retired after a distinguished
career as a trial lawyer. She is a human rights activist and member of
the Moscow Helsinki Watch Group. She is now under indictment on
political charges.

Nina Komarova-Nekipelova, a research pharmacist, is the wife of
the poet and writer Victor Nekipelov, a member of the Moscow Helsinki Watch Group who is serving his second term in camp. Komarova-Nekipelova lives near Moscow.

Lev Kopelev (1912–) fought in World War II, spent many years
in the Stalinist camps, and was rehabilitated in 1956. A writer, German literature critic, and advocate of the human rights movement,
Kopelev was expelled from the Communist Party in 1968, and from
the Soviet Writers Union in 1977. Books of his which have appeared
in the West include: *To Be Preserved Forever* and *The Education
of a True Believer*. Kopelev and his wife, Raisa Orlova, now live in
Cologne.

Vladimir Kornilov (1927–) is a poet, human rights activist, and
former member of the Soviet Writers Union. He is currently a member of the Moscow branch of Amnesty International.

Raisa Lert (1905–) took active part in the publication of the Moscow samizdat magazines *The Twentieth Century* and *Search,* For her participation in the publication of these magazines and for her protests against unfair trials and illegal repressions, she has been expelled from the Communist Party.

Semyon Lipkin (1911–) joined the Writers Union when it was first organized in 1934. He is a veteran of World War II, a poet, and translator of the epics of the Eastern peoples into Russian. Lipkin recently resigned from the Writers Union to protest the expulsion from the union of two young contributors to the collection *Metropol.* He lives in Moscow.

Maksudov (pseudonym) is the author of a study of the population drop in the USSR which appeared in samizdat in the late 1960s. His sociological and demographic articles have appeared in Moscow samizdat journals and abroad. He now lives in the United States.

Anatoly Marchenko (1938–), a blue-collar worker and writer, has spent seventeen years as a political prisoner. He is the author of *My Testimony, From Tarusa to Siberia,* and numerous articles and letters. In September 1981, Marchenko was sentenced to ten more years in labor camp and five years of exile.

Victor Nekrasov (1911–) is a writer and journalist whose first book, *In the Trenches of Stalingrad,* was published in 1946. In 1977 Nekrasov was subjected to harassment by the KGB, the seizure of manuscripts, and detention. He now lives in Paris and is a regular contributor to many Russian-language magazines published abroad.

Raisa Orlova (1918–) is an American-literature specialist who published *The Descendants of Huckleberry Finn,* books about Harriet Beecher Stowe and John Brown, and many articles on American writers while in the Soviet Union. In 1980 she was expelled from the Soviet Writers Union for writing a letter in defense of Andrei Sakharov. She now lives in Cologne with her husband, Lev Kopelev.

Evgeniya Pechuro (1919–), a historian and veteran of World War II, lives in Moscow.

Maria Petrenko-Podyapolskaya (1923–) is a geologist. A longtime participant in the human rights movement, she lives in Moscow.

Grigorii S. Podyapolsky (1926–76), a geophysicist, was the author of a number of scientific papers. He was a member of the Initiative Group for the Defense of Human Rights in the USSR, and took part in the work of the Human Rights Committee, together with Andrei Sakharov.

Grigorii Pomerants (1918–) is a World War II veteran who spent eleven years in camp. A philosopher and expert on Zen Buddhism, his articles regularly appear in Moscow samizdat journals. He lives in Moscow.

Valery Soifer (1936–) is a biologist and author of dozens of books and scientific papers published in the USSR and abroad. He has been subjected to harassment by the authorities since 1979.

Victor Trostnikov (1928–) is a physicist and philosopher who has written several books on physics and mathematics, many published in the West. Since 1975 he has written books and articles on religious and philosophical subjects.

Georgi Vladimov (1931–) is a writer whose books *The Big Ore, Three Minutes of Silence,* and *Faithful Ruslan* have been published in the West. Vladimov, who resigned from the Writers Union in 1977, is chairman of the Moscow group of Amnesty International and lives in Moscow.

Vladimir Voinovich (1932–) is a writer whose books *The Life and Extraordinary Adventures of Private Ivan Chonkin, Pretender to the Throne, The Ivankiad,* and *In Plain Russian* have been published in the West. In 1974 he was expelled from the Soviet Writers

Union. In 1980 he was deprived of his Soviet citizenship. He lives in Munich.

Herbert F. York is professor of physics and director of the Program in Science Technology and Public Affairs at the University of California, San Diego. He was the first chief scientist of the Advanced Research Projects Agency (1958) and the first director of Defense Research and Engineering (1958–61).

Father Sergei Zheludkov (1909–) is a Russian Orthodox priest, a veteran of World War II, and an active participant in the human rights movement. He is a member of Amnesty International and lives in Pskov.

A NOTE ON THE TYPE

The text of this book was set, via computer-driven cathode-ray tube, in a film version of Caledonia, a typeface designed by W(illiam) A(ddison) Dwiggins for the Mergenthaler Linotype Company in 1939. Dwiggins chose to call his new face Caledonia, the Roman name for Scotland, because it was inspired by the Scotch types cast about 1833 by Alexander Wilson & Son, Glasgow type founders. However, there is a calligraphic quality about Caledonia that is totally lacking in the Wilson types. Dwiggins referred to an even earlier typeface for this "liveliness of action"—one cut around 1790 by William Martin for the printer William Bulmer. Caledonia has more weight than the Martin letters, and the bottom finishing strokes (serifs) of the letters are cut straight across, without brackets, to make sharp angles with upright stems, thus giving a "modern face" appearance.

Composed by The Clarinda Company, Clarinda, Iowa

Printed and bound by
The Haddon Craftsmen, Inc., Scranton, Pennsylvania

Typography by Joe Marc Freedman